THE DEVIL'S BREATH

THE DEVIL'S BREATH

THE STORY OF THE HILLCREST MINE DISASTER OF 1914

Steve Hanon

NeWest Press

Library and Archives Canada Cataloguing in Publication

Hanon, Stephen, 1949–
The devil's breath : the story of the Hillcrest Mine Disaster of 1914 /
Stephen Hanon.

Includes bibliographical references. Issued also in electronic format.
ISBN 978-1-927063-29-3

1. Hillcrest Mine Disaster, Hillcrest, Alta., 1914.
2. Coal mine accidents — Alberta — Hillcrest — History — 20th century.
3. Hillcrest (Alta.) — History — 20th century.
I. Title.

TN806.C22A5 2013   971.23'4   C2012-906592-7

Editor for the Board: Don Kerr
Cover and interior design: Natalie Olsen, Kisscut Design
Author photo: R.L. Bolander
Copyeditor: Michael Hingston

Canada Council for the Arts / Conseil des Arts du Canada / Canadian Heritage / Patrimoine canadien

accessCOPYRIGHT FOUNDATION / Alberta Government / The City of Edmonton / edmonton arts council

NeWest Press acknowledges the financial support of the Alberta Multimedia Development Fund and the Edmonton Arts Council for our publishing program. We further acknowledge the financial support of the Government of Canada through the Canada Book Fund (CBF) for our publishing activities. We acknowledge the support of the Canada Council for the Arts which last year invested $24.3 million in writing and publishing throughout Canada.

NEWEST PRESS

#201, 8540–109 Street
Edmonton, Alberta T6G 1E6
780.432.9427
www.newestpress.com

No bison were harmed in the making of this book.

printed and bound in Canada   1 2 3 4 5   14 13

MIX
Paper from responsible sources
FSC® C103214

Dedicated to the memory of Ricardo (Rick) Petrone of the Crowsnest Pass

CONTENTS

"You have seen some of the hands of the miners where they have little blue colourings all through — very fine, I understand — that is caused by a constant picking and the hard pieces of coal flying and injuring them and leaving a certain colour under the skin?"

— a question addressed to Dr. Ross of Hillcrest by the legal counsel for the United Mine Workers of America, District 18, at the coroner's inquest into the Bellevue disaster, 1910

ACKNOWLEDGEMENTS

The details of this book came for the most part from original documents in archives, including the Provincial Archives of Alberta in Edmonton, which holds the transcripts of both the Bellevue and Hillcrest coroner's inquests as well as the transcript of the Hillcrest commission of inquiry; and the Glenbow Archives in Calgary, which maintains a wealth of information on coal mines and mining in Alberta in numerous fonds. In the course of my research I travelled to the Crowsnest Archives in Coleman, Alberta; the Frank Slide Interpretive Centre; the Sir Alexander Galt Archives in Lethbridge; the Fernie and District Historical Society; the B.C. Provincial Archives in Victoria; the University of British Columbia's Special Collections in Vancouver; the Alberta Legal Archives in Calgary; the Calgary Public Library; the University of Calgary libraries; the City of Victoria Archives; and the Victoria Public Library. I owe a huge debt of gratitude to the staff of these facilities, who generously assisted and shared their knowledge. Other facts were gleaned from the National Archives of Canada, both online and by telephone.

Often I have made reference to literature on the subject written by historians to whom I am deeply indebted. Among them I must count first and foremost the Crowsnest Pass historian Ann Spatuk and her team of dedicated workers with the Crowsnest Historical Society, who gathered, wrote, and edited the three volumes of Crowsnest Pass local history, *Crowsnest and Its People,* on which I have leaned heavily. For Crowsnest Pass history largely concerned with the British Columbia side of the provincial border, I recommend *The Forgotten Side of the Border,* edited by Wayne Norton and Naomi Miller. It offers a rich compendium of information about early mining in the Pass. *Alberta's Coal Industry, 1919,* edited by David Jay Bercuson, gives fascinating details of the mines and miners in Alberta. To understand mining in the Pass from a union point of view, I drew upon the worthwhile history of the United Mine Workers of America in Western Canada, Bruce Ramsey's *The Noble Cause.* For information about George Frolick, his experience during and after the disaster, and Ukrainian life in the early days of Hillcrest, the memoirs of his son Stanley, *Between Two Worlds,* edited by Lubomyr Y. Luciuk and Marco Carynnyk, proved invaluable. For CPR history in general, and as it relates to the Crowsnest Pass, I went to *The Canadian Pacific Railway and the Development of Western Canada, 1896–1914* by John A. Eagle. For insight into radical unionism, I reached into the well of information in *F.H. Sherman of Fernie and the UMW of A, 1902–1909: A Study in Western Canadian Labour Radicalism* by Allen C. Seager, which resides in UBC's Special Collections. For coal mining as the miners themselves remember it, and for comparison with Alberta mines, I went to, and greatly enjoyed, Lynne Bowen's award-winning book *Boss Whistle,* which paints striking images of coal miners' lives on Vancouver Island.

Many individuals provided me with priceless help and cooperation, or allowed me the use of valued family photographs. They include Barbara Allan, Patricia Blakely, Hugo Civitarese, Vernon Frolick, Peter Heusdens, Frank Hosek, Jim Hutchison, Mrs. Alice Jamieson, June Kanderka, Belle Kovach, Rick Petrone, Cathy Pisony of the Frank Slide Interpretive Centre, Jean Shafer, Louise Wells, and Dave Welsh. I must also express gratitude to my editor, Don Kerr, and to the people at NeWest for their confidence in this project. During the writing of this book, my friend Rick Petrone of Bellevue passed away. I will always remember how eagerly he shared his knowledge and memories of Hillcrest and Hillcrest Collieries, and the many kindnesses he offered. He is sincerely missed and remembered with fondness.

Although I have gone to many sources for the information required to assemble this history in an effort to compare accounts and eliminate errors, some readers will find mistakes that have eluded me. These are my responsibility, and not those of the historians or others whose work or memories I consulted.

My sincere hope is that this book will give readers a richer understanding of this important aspect of Alberta and Canadian history, and even, perhaps, a small insight into the human heart.

Steve Hanon
Bowen Island, British Columbia
February 2007

HISTORICAL NOTE

I have attempted to provide the first and last names of every figure who appears in this narrative. However, this was not always
possible. In newspapers and official documents of the first half
of the twentieth century, individuals were often referred to only
as "Mr. Smith," or sometimes "Mr. A. Smith." Women in that era
often took on their husband's name —"Mrs. John Smith" or "Mrs.
J. Smith," with their given names left out of documents. In personal practice, first names were used. The result is that the first
names mostly of women but sometimes also of men have been
lost to time, known only to relatives, or hidden in birth and death
records.

THE DEVIL'S BREATH

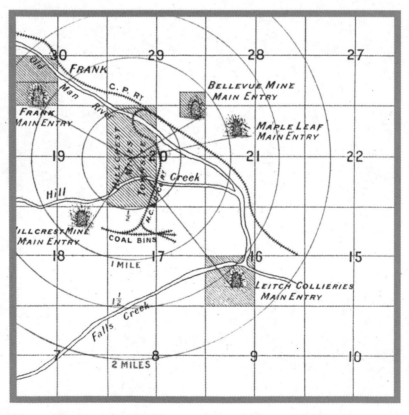

Sketch showing exact location of Townsite of Hillcrest Mines, Alberta.

INTRODUCTION

This book tells the story of a mine explosion in southwestern Alberta in 1914 that killed 189 men. The explosion at the coal mine near the small town of Hillcrest was, and remains, Canada's worst mining disaster and, like all such tragic events, ripped out the hearts of the families and friends of the men killed.

But for many reasons, the events have stood in the shadows of Canadian history. Most Canadians have heard of the Springhill disasters in Nova Scotia in 1956 and 1958, which killed 39 and 74 men respectively, in part because they occurred during the lifetimes of Canadians alive at the time of this writing. Yet many Canadians know little about the Hillcrest events, or even of the extent of coal mining in Western Canada during the first two decades of the 20th century. Central Canadians considered the west a hinterland, noticeable only insofar as it supplied food and raw materials for their bellies and factories. Coal from the west seldom made it east of Manitoba, and very little even into that province. The cost of shipping east could not compete with the cost of shipping west: into Quebec from the mines of Nova Scotia, and into Ontario from Pennsylvania. Western coal mines, and what went on in them, therefore became largely invisible to the population of Central and Eastern Canada, which in fact *was* Canada before 1905, when Alberta and Saskatchewan became provinces. Until then, the plains and mountains of what became Alberta were known only as the Northwest Territories. While the industrial/financial hub of Canada, Montreal and Toronto, supplied most of the capital for western development, including the coal mines, the three westernmost provinces, Alberta, B.C., and Saskatchewan, held little interest among central Canadians. Only those with investments in the West — the romantics who saw it as the repository of their fantasies and hopes, who clung to images, sometimes unblemished by reality, of vast landscapes, and breathtaking mountain scenery peopled with a few noble Indians, and who saw opportunity there — read the few newspaper columns about the region. Among those with eyes and thoughts focused west, the best bought railway tickets and watched the east shrink backwards as they looked with excitement to where they would make their marks. However, coal mines, which did not supply fuel for the mighty manufacturing concerns or the homes of Central and Eastern Canada, were of little concern to the general population, and the Hillcrest disaster was a passing blip overshadowed by the opening act in the theatre of the Great War.

We cannot consider the Hillcrest disaster unique; in fact, disasters in coal mines were all too routine. On the other hand, the very routine of it obscures the meanings of the lives lost. We tend to lose sight of the individual victims of mass death, and of the more serious consequences for a family in that era of the death of a breadwinner. Thus, at its most fundamental, the story of the Hillcrest disaster is the story of the men who worked at Hillcrest Collieries and of their families, whose lives the disaster profoundly and irrevocably altered. Because of the central role it played in their lives, the story of the mine itself, both before and after the disaster, is also worth examination. To tell such a story accurately, to understand what happened and why the key figures did what they did, we must understand the context within which these events unfolded. This context must include the temper of the times: the ideas, attitudes, personal connections, events, and beliefs that shaped their behaviour.

For the most part, North Americans lived until the Great War with a 19th-century attitude of boundless optimism, a belief in the limitless possibilities of human achievement, and the conviction that any individual could succeed, given the self-discipline and work ethic that success demanded. This warm current of thought, which swept into the West with the Enlightenment, was about to clash with determinism, a colder (and older) belief that the individual must be subject to the collective will, and that individual efforts were doomed to failure. The 14-year period from the turn of the century to the Hillcrest disaster and the Great War would see conflicting ideologies become the bloody clashes of armies that would reverberate throughout the 20th century and beyond.

Any effective study of the disaster must also take into account the connected events before and after the explosion, which give us clues to outcomes and motives behind behaviour, as well as what individuals said at the time. We have four sources for what was said: newspaper reports, company records, memoirs, and official documents. Contemporaneous reports, particularly those in newspapers, often lacked accuracy, because of the chaos surrounding the event. It was from newspapers, however, that members of the public obtained their information. While reporters tried to get the facts, the people to whom they directed their questions were preoccupied with taking quick and decisive action, with the best knowledge available, in order to save lives.

Furthermore, the *appearances* of events are often not the *facts* of events. A person may tell a reporter (as well as a court, under oath) what he sincerely believes to be true, but is, in fact, not. Others will deliberately evade or lie, leaving reporters with inaccurate reports. A good reporter pursues truth doggedly, courageously, and may, on occasion, walk a deceptive path to get it.

Finally, this book investigates why the mine exploded, and who, if anyone, caused it. In the study of history, individuals are often discounted as actors, and considered powerless in the face of the large forces and events that mould it. But I have observed that the courage of one often gives courage to others who stand silent, passive, stunned, or frightened into immobility. Within the larger story of the Hillcrest disaster emerge what might seem to be smaller stories of courageous men and women, but which are, in reality, the jigsaw pieces that make up the larger picture. Human weaknesses show their faces here as well, and emerge as other parts of the same puzzle. Both the Hillcrest disaster and the Bellevue disaster four years earlier show men and women at their best and, in some cases, their worst. The "cause" of the Hillcrest disaster remains complex and wrapped in a figurative smoke that may never be penetrated. The elusiveness of the cause, however, only contributes to the power of this tale of passion, death, survival, and, yes, love. In the search for truth, one may be humbled by the mysteries, but concurrently awed by the discoveries of human resilience and determination in the battle against great odds. In the story of the Hillcrest disaster lies part of the story of each of us. Read on, and know thyself.

CHAPTER 1
THE LURKING THREAT

You are an experienced miner, 25 years old, not a tall man, but strong. You are an immigrant to Canada, one among the many Ukrainians who work in the mine, but you plan to return one day to your homeland with the money you've earned here. Still a bachelor, you send a portion of your earnings back to your father in the Ukraine. At this moment, you are deep in the guts of the Hillcrest Collieries mine in the Crowsnest Pass of southwestern Alberta. The year is 1914; the date, June 19. Your name is George Frolick.

You have just delivered eight carts laden with high-quality bituminous coal to the no. 2 slant, from which a wire rope attached to a winch will haul the cars to the surface. After you attached your carts to the hoist, you used the telephone to tell the hoist operator that another load was ready to be brought to the surface. The only illumination comes from the Wolf safety lamp you carry. The lamp casts a dim light that barely penetrates the darkness of the hole in which you work, and the flame of the lamp will not ignite the explosive firedamp gas released from the coal because of the lamp's wire mesh barrier. That is a very good thing.

The real danger lies at the face of the mine, where the miners extract coal. The coal exudes methane, which drifts and collects in pockets in the mine's roof. An open flame would set it off, but this is unthinkable. The explosion would mean certain death.

Now you are ready to haul a second load to the slant. You work in the dim lamplight with other Ukrainians. You understand each other, help each other, support each other. Alone as you lead the horse, your lamp exposes the wooden timbers that support the roof and protect you from the millions of tons of rock above. You do not think about the weight that presses down from above. You focus on where you step in the constantly moving shadows. As you marshal your horse into position to haul the second of eight loads to the slant, something happens — perhaps the result of someone's miscalculation, a human error, the end result of a chain of circumstances, unforeseen, unexpected, harmless, no doubt, in another context, perhaps the result of human action, perhaps not, perhaps nature, or, some might say, the work of the Devil. Then again, anything that happens in the mine is the result of human action, because humans have opened the Pandora's box, burrowing into the earth, working the coal. Now the responsibility is theirs. In any case, it happens, close to the face where men pick away at the coal to loosen it, so they can shovel it into the chute, where it slides down to the cars that you drive to the slant. You don't even hear it, as other men do in distant parts of the mine, although the origin is not far from where you now walk. The tremendous force of the concussion knocks you to the floor of the level, senseless. You lie next to the steel tracks, your face on the grit of granular coal and dust. You hear nothing, you see nothing, but through chance or luck, somehow you still live, you still breathe. You are unconscious and helpless, but you live. For the moment.

And then you awaken, confused, and disoriented in the pitch black. You wonder for a moment if you are blind. What has happened? Your head is pounding. You hear the sounds of men, but not the usual sounds. These voices tear raggedly from the throat, from some deep, bloody gash, the exposed animal part of men, high-pitched: the cries of men in agony, bleeding and dying, and those who can cry out for salvation.

What nightmare is this? My God, what nightmare is this? In an instant,
you decide not to die. You accept responsibility for life. From this point
on, whether you live or die is entirely up to you. This is no longer about
coal mining, or anything else remotely civilized; it is about survival, the
will to live, to walk or crawl towards the light, and into it, and to suck
in the fresh air, suck in the fresh air, and to see the sky again.

The men who ventured underground to dig coal entered a realm of
darkness where, myth would have it, only the spirits of the dead
dwelled. In fact, danger oozed from the walls and permeated the
very air that the coal miners breathed, and when a careless or
inattentive man walked into the mine, he walked to his grave. It
must be said that even the most safety-conscious of miners could
fall victim to the dangerous nature of the job, or to the mistake of
another miner. The mine environment, particularly in 1914, was
full of all manner of danger: rock bursts, bumps, explosive gas,
poisonous gas, suffocation, rock falls or caves, heavy machinery, equipment failure, explosives used in loosening coal, horses,
mules, sudden flooding. All of these dangers could be managed
effectively with research, experience, realistic appraisals, precautions, cooperation, and individual responsibility — but they
could never be eliminated entirely. In an underground coal mine
in 1914, death was a constant companion, and the fatality record
speaks for itself.

The list of fatal accidents at Hillcrest over the years reflects the
ever-present danger in underground mines in general. On January
29, 1910, William Cheryk died in the Hillcrest mine after he was
crushed between two mine cars. No one was blamed. That same
year, falling rock fatally crushed a miner named George Martin as
he and two other men repaired a chute. In 1912, Samuel L. Wilson
died when he was overcome by gas exposure. In December 1913,
31-year-old Frank Rose died when he was crushed by falling
rock. An inquest before Coroner Pinkney, with District Mine

Inspector Scott of Edmonton present, concluded that the death was accidental.

After the disaster of 1914, accidents continued — some fatal, some not. On November 20, 1914, an accident killed two men. The hoist cable snagged and broke, and Pacifico Cimetta, a rope rider, and William George Thomas, a horse driver, were struck by, or thrown from, the loaded cars that rolled backwards and out of control down the slant. In a similar accident in May 1915, runaway cars on the slant struck the car next to miner George Frolick. His leg was crushed. On July 10, 1923, a piece of the main roof fell on 33-year-old miner Earl L. Eckmier, crushing his head. In many of these accidents, blame could not be attached to any one person. In others, a moment of inattention or carelessness on the part of the miner was responsible. In all, over the course of the working life of the Hillcrest mine, 228 men were killed.

An accident at a mine in Arkansas provides a perfect example of the horrors faced by miners in underground coal mines, as described in this newspaper report of November 18, 1904:

> Ventin Abrams, an Italian miner employed in Mine No 8, of the Bolen-Darnall Coal Co, was killed while at work a short time before noon Wednesday. The first intimation he had of danger was when a large rock fell upon his foot, throwing him against and partially over the car which he was loading. Another miner, who was working with Abrams, was also struck by the rock, but he escaped without injury. Seeing that tons of rock would come down in a few seconds he, with several others who were working near, made for a place of safety, not daring to go to the rescue of Abrams, as to do so would be to invite certain death. Abrams knew more rock would fall and appealed piteously to his comrades to come to his rescue, at the same time making superhuman efforts to free himself from the rock that was holding him. In a few seconds the rest of the rock came down, crushing the life

out of him. Death was instantaneous. Deceased was about 35 years of age and leaves a wife and two children in Italy. The remains were taken to Moberly for interment. This is the second fatality which has occurred in this mine, which has been sunk about four years.

Mine disasters in which large numbers of men were killed grabbed newspaper headlines and shocked the public, but the greatest danger to miners came from falling rock or coal from the mines' roofs. In 1911 in the United States, such falls "killed over three times as many miners as were killed by explosions, and as many as were killed by all other accidents underground."[1] U.S. mining official George S. Rice says these deaths from falls occurred in widely scattered mines, were not reported except by local newspapers, and therefore went largely unnoticed by the public at large. Yet these accidents claimed the lives of a huge number of men: the death toll in the U.S. came to a staggering 1,321 in 1911 alone. Mine car and locomotive accidents killed an additional 438 men that same year. Other dangers, however, also threatened the coal miner.

The assistant general manager and chief engineer at Hillcrest, William Hutchison, provides an illustration of one particular threat in some deep mines, where the depth of cover exceeds 400 feet: a bump.[2] Bumps occur randomly when the pressure created by the mass of rock above a mine tunnel creates high stresses on the pillars, which then transmit the pressure to the underlying pavement. In Hillcrest the strata would often heave up from the floor, rather than, as in some mines, burst out from the pillars. The bump, then, is the result of a redistribution of pressures altered by the mining process of tunneling.

Bill and his brother David, the mine's surveyor, experienced an extreme bump in 1927. Among their responsibilities was providing a weekly progress report for the mine's owners in Montreal. This report included measurements of the amount of air travelling in the various splits compiled from anemometer readings and, at that particular time, measurement of the water level in an abandoned slope and some connected levels deep in the mine. The men had taken the measurements, and were well up the abandoned slope when a violent bump shook it. The extreme violence of the upthrust, powered by millions of tonnes of rock, shattered the support timbers and twisted the rails as it threw both men into the right rib. Their helmets were blown off. Fortunately, the bump left the lights on the helmets intact and functional, but the displacement kicked up coal dust so thick that the men could barely see or breathe, and they nearly choked to death. They managed to crawl up over the brow, which was nearly to the ceiling, and down onto the level, badly shaken but no worse for wear. It could have killed them.

Mine disasters, however, which are usually caused by explosions of methane and coal dust, are an entirely different matter. The first recorded mine disaster in Canada occurred at the Drummond Mine in Westville, Nova Scotia in 1873, where sixty men died in an explosion. It was the first of many disasters that left families devastated, the public puzzled, and mining officials at a loss. The first major disaster in Western Canada occurred in 1887, when an explosion in Nanaimo, B.C. killed 150 men. Another explosion killed seventy-five men at the nearby Wellington mine the following year. In 1891, 125 died in Springhill, Nova Scotia. In 1901, sixty-four died in an explosion in Comox, B.C. In 1902, at Coal Creek, near Fernie, B.C., an explosion killed 128 men. In 1903, 128 died in an explosion in Reserve, Nova Scotia. These disasters are among the largest in Canada, but hundreds of men died in other disasters, some large, some small. Other men died in smaller numbers. Between 1904 and 1963, 1,257 men were killed in Alberta coal mines alone.

In terms of coal mine disasters on the world stage, however, and the sheer numbers killed, Canada's fatalities are modest. The worst known coal mine disaster is believed to have occurred during the Second World War in Manchuria. On April 26, 1942, 1,549 men died in a disaster at a time when the Imperial Japanese Army occupied Manchuria, and treated the Chinese as slave labour. Another disaster took the lives of 1,100 men in a coal dust explosion in Courrières, France on March 10, 1906. In that disaster, rescuers brought out thirteen men who had been trapped in the mine for twenty days. The worldwide toll from major disasters alone is truly staggering. In Britain, from 1851–1940 8,331 men lost their lives in major mining disasters. The day-to-day toll was much higher. In just ten years, from 1875–1885, 12,315 people died, including women and children. In China, major mining accidents continue to this day. In 2005, the Chinese government reported that it would close 7,000 coal mines (of about 24,000 in total) in a safety crackdown on the accident-plagued industry. The Chinese government reported that 5,000 Chinese coal miners had lost their lives in 2004 due to lack of fire control and proper ventilation, or failure to enforce safety rules. According to statistics from China's State Administration of Work Safety, China produced thirty-five percent of the world's coal in 2011, but reported eighty percent of the total deaths in coal mine accidents.

The worst disaster in the United States wiped out the lives of 362 men on December 6, 1907 in Monongah, West Virginia. This disaster is claimed by some to have led to the first Father's Day. If the Hillcrest disaster had taken place in the U.S., it would be the seventh-deadliest mining disaster in that country's history. In Canada, it ranks first.

Account must be taken, of course, of the total numbers of men who worked in these coal mines. From here we can determine a more accurate view of the danger of coal mining. In 1900 alone, 448,000 men worked in coal mines in the United States. Between 1891 and 1900, 1,473 men died in mine explosions.[3] Just the major explosions in the year 1900 killed 276 men. Yet these numbers do not account for deaths in U.S. mines from other causes. The Cherry Illinois Disaster of November 13, 1909, for example, took the lives of 259 men after a bale of hay caught fire and spread.

Men were burned alive or suffocated, and rescuers perished in their attempts to save fellow miners.

In Alberta, the 121 mines that operated during 1909 employed a total of 5,207 men. Seven men died inside the mines, and two outside. From 1905–1909, the amount of coal mined increased from 811,228 tons in 1905 to 2,174,329 tons in 1909[4] — an increase of 268 percent. The number of tons of coal mined per fatal accident during 1909, then, was 241,592. Put another way, it was one fatality for every 579 coal miners. These numbers suggest that the greater fatality rate in the U.S. was due to larger mines, where a greater number of men were put at risk in any possible explosion.

In 1912, 6,661 miners in 243 mines in Alberta produced 3,446,349 tons of coal, while 21 men were killed.[5] For that year, one man died for every 317 miners working, or for every 164,112 tons of coal produced in the province. As the mines in the pass grew larger and deeper, with more gas, men, and machinery, the numbers of fatalities approached those of pre-1900 U.S. fatalities.

In 1915, the total output grew to 3,360,818 tons of coal, and by the end of 1920 to 6,908,923 tons.

Even a small mistake by a fellow miner can nullify every precaution taken by reasonable, responsible, experienced men to protect themselves. Carelessness with equipment or explosives, an open light, or inadequate ventilation can kill or injure others, as well as the man or men responsible. The only way to completely eliminate mine accidents would have been to cease the operation of underground mines altogether — a practical absurdity, given the massive importance of coal as an energy source throughout the 19th and at the beginning of the 20th centuries.

Coal was to the economy, then, what oil and natural gas are today. In fact, coal had been important as a source of heat and energy in Britain since the late 1500s, when the population had exploded, farmers had largely razed their forests to make way for agriculture, and wood had vanished as a source of large-scale fuel consumption. Among Western nations, Britain stepped into the coal age first. But all this happened at a time when infants were still dying in large numbers, fatal diseases cut wide swaths through the population, and medical treatment could be described as rudimentary at best. During the early 1600s in England, for example, life expectancy was only about 35 years, largely because two-thirds of all children died before the age of four. This began to change when food production increased, as methods of agriculture improved.

Coal fuelled the Industrial Revolution, which began in the late 1700s. The urgency to get coal out of the ground and the need for men to find employment to feed their families sometimes led to unsafe working conditions — and this attitude certainly contributed to accidents in an era when far less importance was attached to human lives. A general lack of technical and scientific knowledge also hindered the improvement of safety standards and equipment. As the Industrial Revolution advanced, and coal became even more critical as a source of energy, so too would scientific and technical knowledge evolve, and, consequently, safety within the mines.

The men who took up jobs as miners knew full well the dangers involved, although often mine management and miners alike dismissed the danger in particular mines:

Correction [of poor safety practices] was difficult to bring about, because of the settled practices and deep-rooted habits in which both mine officials and mine workers persisted. The truth that untrained and for the most part poorly educated men could not safely carry on the work in most of the mines without close supervision by well-trained and alert foremen after all hazardous conditions were carefully controlled was recognized by many. Only a small number of inspectors and other officials and engineers believed that correction was possible. By far the greatest danger was refusal of the great majority of officials and workers to admit that explosions in their mines were more than a remote possibility.[6]

It is, it would appear, human nature to slide into the lazy thinking that claims "It can't happen to me" or "It can't happen here." On the other hand, the easy, smug criticism often levelled at mine management by politically motivated writers fails to take into account the context of coal mining at the beginning of the 20th century. Many miners and managers, for example, were not particularly well-educated. In fact, this was true of labourers in general at the time. Men who failed to understand the dangers they faced could not be expected to practice safe working habits. At the same time, even a door left open in a coal mine can interrupt or nullify the ventilation plan designed to remove poison, explosive gas, or foul air. Such mistakes take only a second of inattention, and the payoff can be death. In 1919, at a hearing of the Alberta Coal Mining Industry Commission, the provincial mines inspector for the Drumheller district told of men in the mines who "uncontrols [sic] the system of ventilation by leaving doors open. That's what happens very often." He added that as inspector he had to explain to the men that the doors were put in for the purpose of ventilation.[7] Both common sense and evidence tell us that safety in underground coal mines depends upon relevant knowledge by miners and mine management.

Hiram Humphrey, a coal mining expert with the US Bureau of Mines, and the author of *Historical Summary of Coal-Mine Explosions in the United States, 1810–1956*, referring to U.S. coal mine disasters, wrote:

> [T]he published reports of state mining departments show that the companies ordinarily claimed that all provisions of the mining laws were being complied with and that disasters were wholly due to illegal or careless acts of workers, for which the company was not responsible. Coroners' juries ordinarily brought in a verdict that no blame could be attached to anyone for the explosion. State mining inspectors usually reported that their investigations showed that violations of mining laws had not been evident during inspections before the explosions.[8]

It should be added that miners and their unions also denied any individual responsibility for accidents, and tended to blame the mining companies. But as U.S. mining official J.A. Holmes says in his preface to a miners' circular published in 1912, "[L]ittle is to be gained by declaring that the miners or that the mine owner is mostly to blame. As a matter of fact, neither the mine owner nor the miner acting alone can entirely prevent such accidents. Each has his part to do: each is responsible for doing his part well."[9] The only way to save lives was through cooperation. Certainly the prevention of explosions demanded it.

Two gases threaten the lives of miners in underground workings: methane (CH_4, also known as marsh gas, or firedamp) and carbon monoxide (or afterdamp). Firedamp occurs naturally in coal seams and, because of its lightness, tends to collect next to the mines' roofs. The coal generally discharges methane at a regular rate, but sometimes it does so with large quantities under pressure, due to fissures. Every mine is different. In some mines, the pressure can reach 400–500 lbs/in^2 and release with violent force. Colourless, tasteless, and odourless — but not poisonous — methane can explode only in the presence of sufficient air, and reaches maximum explosive force when the ratio of air to methane is 9.5:1. At this ratio, all of the constituents of the gas are consumed. When methane explodes with a smaller percentage of air present, the incomplete combustion of the firedamp produces carbon monoxide (CO), the miner's second deadly enemy. The incomplete combustion of coal dust in a coal dust explosion also produces CO, which offers the miner no clue to its presence, as it has neither colour, taste, nor smell. Unlike methane, however, it is extremely poisonous. When inhaled, a very slight amount results in giddiness, trembling, weakness, and severe headache, then fainting. A one percent mixture is sufficient to cause death. Also present in afterdamp will be a certain amount of carbon dioxide (CO_2). In the absence of mine explosions, CO, generally speaking, is not an issue. The primary safety goal in coal mines is, therefore, twofold: the elimination of explosive gas by means of efficient ventilation, and the removal of any means of ignition of explosive gas.

Coal dust also presents a deadly issue for miners. Over the long term, it damages the lungs of those who breathe it, leading to an often fatal condition known as pneumoconiosis, or black lung. More immediately, coal dust is also an explosive threat in coal mines and anywhere else that coal is handled, such as tipples or boiler rooms. It presents little threat when settled on the floor or support structures, but when suspended in the air, as it does following a methane explosion in an underground mine, it creates a powerful risk of secondary explosions. In fact, most firedamp explosions also involve coal dust explosions. A shockwave ahead of a methane explosion will raise coal dust from the floor and wooden support beams, and, once in the air, it will get ignited by the prior explosion, and so on. Such explosions become self-fuelling. Mine management began to mitigate the dangers of coal dust with finely ground non-inflammable rock dust, usually limestone, which was sprayed on walls, supports, and floors.

Historically, in the United States as well as in Canada, the causes of explosions were clear. "The neglect of ventilation, omission of testing for gas, use of open lights in gassy mines, tolerance of smoking, use of black powder, overloading shots, and failure to remove coal dust were factors often reported." [10] Humphrey adds that the main source of ignition listed in reports was open lights or smoking. A minor source was blasting.

Miners blasted or "fired shots" to loosen the coal. At the face, miners would first undercut the coal at the bottom, drill boreholes for blasting, and then summon the fire boss to set the charge and blast to loosen the coal. Blasting made it easier and faster for the miners to load coal. Mine managers, on the other hand, often preferred that miners not fire shots, because of the danger involved in the ignition of coal dust and firedamp. Miners at Hillcrest sometimes complained that they were not allowed to fire more shots.

In the early days of mining in Alberta, miners usually took their own explosives into the mine, and fired the shots themselves. An incident in the early days of mining in the Crowsnest Pass reveals the danger of explosives in the hands of miners. A bachelor invited some friends to his shack for dinner, but just as they entered, an explosion blew them out of the house and splattered food everywhere. Luckily, no one was seriously injured, and they had a good laugh about it. To protect dynamite from water drips in the leaky, poorly built shacks, miners often stored their dynamite in ovens. The host had forgotten to remove his explosives before he started his fire in the stove.[11]

The wife of one Crowsnest miner recalled that her husband kept dynamite that he shouldn't have had. In the morning he would take some of it with him to work to help bring down coal. When she asked him if it was safe, he replied, "Oh, everybody else does it."[12]

At the same time, at the face, where any open flame invited disaster, miners lit their fuses with matches. Not only were matches dangerous, but black powder also proved itself a dangerous explosive in coal mines because of the possibility that it might set off a coal dust explosion. Eventually, fire bosses took over the shot firing and began to use electrical ignition instead of matches. Mining law later made it illegal for anyone to take pipes, tobacco, or matches into a mine. Dynamite was used in the Hillcrest Collieries as late as 1909, but by 1911 the fire bosses used an explosive called Monobel, which was not supposed to create a flame during detonation. This was in contrast to blasting in the United States at the time, when black powder was still being used. In the Crowsnest Pass, the collieries sold Monobel to miners by union agreement, but it was never discharged by the miners themselves. At Hillcrest, fire bosses ignited Monobel with an electrical ignition system of batteries and wires. As a result, blasting in Hillcrest became much safer than it had been in the past.

One could argue that the discovery that led to the greatest improvement in mine safety was the practical mine safety lamp. Until the widespread use of safety lamps, miners used open flames produced by oil lamps, despite the clear danger from the possible ignition of firedamp or methane. Miners attached the lamps to their cloth caps with a clip or hook. One early mining practice was the appointment of a man to go into the mines with a candle to ignite methane deliberately, to get rid of it before it accumulated in large amounts. This was the origin of the term "fire boss." Unfortunately, many of these men had no idea if the gas had already accumulated in large amounts, and often their lives ended prematurely. The use of fans instead of combustion to eliminate explosive gas proved to be a development nearly as important as the safety lamp itself.

Four men contributed to the development of the first safety lamps, which varied in practicality: a German, Alexander Von Humboldt, whose ideas seem to have fallen by the wayside, and three Britons, Sir Humphrey Davy, George Stephenson, and Dr. William Clanny. The Britons worked on their lamps at the same time, and knew of each other's work. But none patented his invention, and each improved upon the ideas of the others. The urgency to produce a safe means of lighting for miners became a public issue after a mine explosion in England in 1812, which was caused by candles that ignited methane and/or coal dust. The deaths of these ninety-two men and boys sparked the formation of a society to study and prevent mine explosions, which eventually approached Davy for help. His efforts led to the discovery that wire gauze would cool a flame or burning gas on one side of the gauze, to the point that it would prevent ignition of explosive gas on the other side. Stephenson also invented a safety lamp in 1815. The third inventor, Clanny, published a paper in 1813 describing his invention of a safety lamp whose practicality proved limited until he incorporated the concept of the wire gauze. In the end, each man claimed that he was the inventor of the first safety lamp, but the work of all of them contributed important elements to safety lamp design, which saved many lives until they were largely discontinued in the 1930s.

At Hillcrest, and generally throughout the Crowsnest Pass, companies provided miners with Wolf safety lamps, manufactured in England and based on a German design. They had proven themselves rugged, reliable, and inexpensive. Each had double gauze and a bonnet for extra protection from gas under pressure, which could nullify the protection of a single gauze. Perhaps most importantly, the Wolf lamps were what was known as self-lighting — they were lit without a match, by a system very like a modern lighter. This meant that there was no need for anyone to carry matches into the mine anymore.

While the Wolf safety lamp protected miners from the accidental ignition of methane, equally important was its use as a tool to test for the presence of firedamp. If a lamp was introduced into an atmosphere containing firedamp, the flame became elongated, with a blue cap of burning gas over the top of the flame. If a miner became accustomed to a certain height of flame for light, he could detect the presence of a small percentage of gas; with greater amounts of methane the flame became longer and longer. An almost pure methane atmosphere would extinguish the flame completely because of the lack of oxygen necessary to fuel it.

It is difficult to explain the sensation of absolute darkness in a coal mine to someone who has not experienced it firsthand. Once a light is extinguished in a demonstration, the darkness becomes, for many non-miners, unnerving to the point of intolerability. Most people will wonder why a man would choose the life of a miner when faced with the darkness, the danger, and the fear-inducing thought of the weight of rock above. Only the miner and his family understood the strong sense of camaraderie among miners, of "being one of the boys." Contrary to popular misconception, many miners loved their work and the life it gave them, and would not have chosen any other profession. When work was done, they knew how to laugh and enjoy life. Characteristically, and perhaps above all else, miners cultivated a sense of loyalty to one another that grew in part from their reliance on each other (for rescue in the event of a mishap), their reliance on other mining families (during hard times), and perhaps the shared psychology of men who chose to work beneath the earth. Miners would work fearlessly and tirelessly to rescue co-workers caught in perilous circumstances in the dark tunnels of the mines.

However, this willingness to rescue trapped fellow miners, even those who might be complete strangers, was not enough. Efficient rescue efforts required the most up-to-date technology available. The Bellevue disaster in 1910 alerted Alberta mining officials to the importance of properly trained and equipped mine rescue teams that could save lives after underground explosions, when poisonous gases became the major threat to life. In the absence of an Alberta mine rescue team, the newly established British Columbia mine rescue team trained and equipped with Draeger breathing apparatus rushed from British Columbia to Bellevue in Alberta, and rescued a number of men trapped underground in an atmosphere of deadly afterdamp. In the face of such a clear demonstration, the Alberta government equipped its own mine rescue car, and trained a superintendent to look after it. That first car operated between Lundbreck and Coleman. In addition, a mine rescue station was set up at Hardieville near Lethbridge, with substations at Commerce, Coalhurst, and Taber. As part of the initiative, the superintendent of the mine rescue car assumed responsibility for training, held courses in mine rescue, and issued certificates to those who passed.[13] Teams equipped with rescue breathing apparatus and pulmotors were no guarantee of rescue after an accident, but they increased the odds in favour of survival. And that, for the men underground, was good enough.

CHAPTER 2
LIFE AND IDEAS IN THE PASS

The Kootenays lie at the west end of the Crowsnest Pass, the southernmost of the Canadian passes that slash through the Rockies from the foothills of Alberta to the interior of British Columbia. Prospecting, mining, and smelting in the B.C. Kootenays and south of the 49th parallel eventually gave birth to the coal mines on the Alberta side of the Crowsnest Pass. Activity in the Kootenays began in 1864, when miners bored holes into the rock in pursuit of the glitter and promise of gold, which had been discovered on Wild Horse Creek, near Fort Steele. The strike drew hundreds of men to the area, but by 1865, a new gold strike on the Columbia River largely emptied the towns that had popped up around Wild Horse Creek, although prospectors still wandered the area, and mining continued. The get-rich-quick gold miners adopted an itinerant lifestyle: stake first or strike out. One of the men drawn by gold, William Fernie, left England as a boy, in a search that took him to Australia, South America, through the United States, and up into Canada to Wild Horse Creek. Yet he ultimately became not a gold miner but the gold commissioner for the B.C. government. And Fernie would build his wealth not from gold, but in part from coal. Ironically, a gold prospector by the name of Michael Phillips and his companions discovered coal spilling out of the ground in 1874, but for them it held no interest. The timing was wrong, with neither to markets to buy it, nor the transportation to move it. Those would both come later. Over the next twenty-five years, the impetus for the development of coal mines came largely from south of the border, where large hard rock mines and smelters consumed vast quantities of coal.

In the 1880s, Fernie began to see coal as an opportunity. He gathered investment and hired crews to prospect and stake coal deposits, which by that time had become common knowledge. With his partner James Baker, the men arranged financial backing from a group of Victoria businessmen and formed the Crow's Nest and Kootenay Lake Railway Company. Among other talents, Baker brought his connections with men of influence to the partnership with Fernie. Baker had served in the British navy as a teenager, joined the army and fought in the Crimean War, earned a degree from Cambridge, and served as advisor to Prince Albert and Benjamin Disraeli. He was elected to the B.C. Legislature in 1886 only two years after his arrival in the province. Baker's prominence meant that he could approach other influential men. He discovered, however, that the Canadian Pacific Railway (CPR) had no interest in his coal property. During a trip to Ottawa in 1891, he tried to arrange a federal subsidy to build a railway, the British Columbia Southern, to his coal fields. The CPR's William Cornelius Van Horne dismissed the matter, and even ridiculed Baker to the prime minister. Baker's efforts to interest James J. Hill, the tycoon of the Great Northern Railway, also failed.

Only when the mining industry in southeastern B.C. began to boom in the mid-1890s did the CPR recognize the opportunity that Fernie and Baker had identified more than a decade earlier. The mining boom in the Kootenays changed the entire picture for all of the competing interests. Hat in hand, Thomas Shaughnessy shuffled off to Ottawa and begged for a subsidy to build the railway, appealing to Ottawa's fear of the United States. Raising the spectre of so-called "bad America," J.J. Hill's railway incursions into southern B.C., as well as the looming threat of America's Manifest Destiny, worked. The CPR got federal assistance, and as a result, Fernie and Baker got what they wanted, too: $100,000 from the CPR for the purchase of their charter for the British Columbia Southern Railway, as well as heavy investment in the Kootenay Coal Company by business interests in Toronto. This company became the Crow's Nest Pass Coal Company, a major player in the future of the Crowsnest Pass. The tripartite agreement was complicated, but with the purchase of the railway charter by the CPR, the Crow's Nest Pass Coal Company acquired a land grant of 20,000 acres per mile (a total of 3.75 million acres) and six square miles of coal lands near the Pass. At the same time, the CPR surrendered 250,000 acres of coal lands. As for the assistance given to the CPR, the federal government granted a cash subsidy of $11,000 per mile, for a total of $3,404,720 — almost thirty-five percent of the line's total construction cost. The agreement had some strings attached, however. The CPR had to agree not to mine coal in the area for ten years after the construction of the line, but the CPR saw the opportunity offered by the proposed agreement: the massive land grant, the railway charter, and the coal lands. The CPR signed. The deal, as well as its outcome — the Crow's Nest Pass line, from Lethbridge, Alberta, to Nelson, B.C. — became the spark that would ignite the development of Crowsnest Pass coal mines.

Formal portraits of Fernie and Baker taken around this time show two prosperous men at the height of their powers. In Fernie's face we see intelligence coupled with a fierce energy. He stares with challenge at the camera. In Baker, however, we see a more subdued man of quiet strength and confidence. He looks down, his face in shadow, at the book in his hands, clearly a symbol of his dedication to learning and knowledge. It was these two visionaries, perhaps more than any other individuals, who understood the potential of the Crowsnest Pass and advocated its development with so much passion. They paved the way for other men of vision, such as Charles Plummer Hill, who we will meet again in the next chapter, as did the CPR. When the CPR constructed the Crowsnest line in 1897 and 1898, it opened up the entire pass, and the Kootenays, to development.[1]

Coal mining in the Crowsnest Pass largely centred on two competing railways in southern B.C.: the Canadian Pacific Railway, and the Great Northern. There is a certain irony in it. Two men of American birth ran the CPR: Cornelius Van Horne and Thomas Shaughnessy. Whereas James J. Hill (no relation to C.P. Hill), a Canadian, ran its American competition in the Kootenays. Hill had been a member of the original CPR syndicate, which was formed to build the Canadian transcontinental railway, and a member of the CPR board of directors. It was he who had invited Van Horne to come to Canada to join the CPR in the first place. When the CPR decided on an all-Canadian route, instead of the route through the United States desired by Hill, he left Canada and the CPR for good. Instead, Hill built his own railway, the Great Northern, which entered into fierce competition with the CPR in southern British Columbia. By 1901, the year Van Horne vacated his presidency of the CPR, J.J. Hill had acquired thirty percent of the stock in the Crow's Nest Pass Coal Company, which now supplied coal and coke[2] from its mines at Coal Creek, Michel, and Morrissey not only to the CPR smelters at Trail and Nelson and the Granby and Greenwood copper smelters, but also increasingly to U.S. smelters at Great Falls, Montana. By 1906, Hill had acquired a controlling interest in the coal company. The CPR's new president Shaughnessy became convinced that Great Northern gave U.S. smelters a price break on coke shipments that it would not give on coke shipments to the CPR-owned Canadian smelters. Publicly, he accused Hill's company of violation of the tripartite agreement of 1897. This led Shaughnessy to conclude that the

CPR had to find or develop new sources of coal to supply its B.C. smelting operations if it were to remain competitive.[3]

As manager of the CPR smelter at Trail, William H. Aldridge had earned the trust of Shaughnessy, who on July 4, 1903 put him in charge of the company's mining and prospecting operations. Aldridge had succeeded brilliantly when he implemented a new method of lead extraction and built a new lead refinery at the Trail smelter. Coal, however, was a different matter. Aldridge headed his first coal prospecting and development project at Bankhead, just north of Banff, but also sent a party to prospect for coal in the Crowsnest region. In June 1902, that party discovered large seams near Hosmer Station, but the property was tied up in the 1897 agreement with the Crow's Nest Pass Coal Company. In 1904, during development at Bankhead, Aldridge wrote a letter to Shaughnessy, outlining the CPR's coal prospects. He wrote in the letter that "the Bankhead property near Banff is the only one between Revelstoke and Medicine Hat on the Main Line which is likely to be a producer of first class domestic and steam fuel."

The CPR had been late in its exploration and development of suitable coal properties. This was a tactical mistake on the part of Van Horne and Shaughnessy. Another company, the McNeill Company alongside Sir Sanford Fleming, controlled the rest of the Bow Valley between Banff and Canmore, except for one square mile northeast of Canmore. The coal lands available east of the gap in the Bow Valley produced only low-quality lignite, completely unsuitable for locomotives. Aldridge pointed out that even Bankhead's semi-anthracite coal might require a special type of locomotive. He argued that Bankhead could only be profitable in the absence of high-quality anthracite if the CPR used locomotives designed for that type of coal.

Aldridge also pointed out that the Hosmer property in B.C., although valuable, had been tied up in the agreement with the Crow's Nest Pass Coal Company for some time. Coal fields on the Elk River, although promising, would require the construction of forty miles of railway, a prohibitively expensive and complex undertaking. The McInnes property, he wrote, also presented problems. It would require nine miles of railway, and was a small property. The coke was better than that produced in the Northwest Territories (now Alberta) at Lille, in their Belgian ovens, but contained twenty percent ash, and would have to be picked to remove slate. All said, it would be another expensive development with only very limited prospects for success.

East of the mountains, on the Crow's Nest line, the CPR had been unable to secure any coal property for various reasons, including "price, unsatisfactory title, inaccessibility, or questionable condition of the measure and coal." Aldridge suggested that because of the difficulty to determine whether any of the operations at Coleman, Blairmore, Lille, Bellevue, and Frank would be able to produce suitable coal, the best policy would be to encourage those developments, and then sign contracts with those who were able to furnish the best coal. The only additional suggestions he could make were to continue exploration and development work where prospects seemed promising, and to try to secure a coal property that produced coal that was at least as good as what they were already getting from the Galt mine in Lethbridge. Galt coal was not the high-quality bituminous coal they wanted, but the CPR used large quantities of it anyway. In the absence of anything better, implied Aldridge, it would have to do. In 1912, the CPR began a long-term lease of the Alberta Railway and Irrigation Company, which owned the Galt mine after the CPR purchased a majority interest in the Lethbridge company. It was the only coal mine investment ever made by the CPR that proved profitable. The CPR still faced a serious quandary, however: it didn't have the quality of coal it needed for stationary boilers, locomotives, or the production of coke, and the coal that they did have cost them serious money because of its lower efficiency.

After the $2 million expenditure by the CPR towards the development of the mine and briquette manufacturing plant at

Bankhead, the Pacific Coal Company mine went into production in 1905. But the coal soon proved to be poor quality. The CPR finally shut down the mine in 1923. The development had been based on desperation and hope — a poor substitute for planning and knowledge.

The second coal development under Aldridge's direction began production in 1908 after another expenditure of $2 million to build a large surface plant and coke ovens at Hosmer, just east of Fernie. It also failed, almost from the start. The badly faulted seams hiked production costs to a level that was impossible to sustain. In the end, the Bankhead and Hosmer mines cost the CPR approximately $4.5 million. Shaughnessy was naturally displeased with the wasted expenditures, and criticized Aldridge for it. The CPR's two ventures into ownership of coal mines had failed miserably. In February 1911, Aldridge resigned as managing director of Consolidated Mining and Smelting to take a position in the United States.

A parade of colourful characters crowds the history of the Crowsnest Pass from the period before the railroad through the 1920s. Native buffalo hunters, gold seekers, traders, explorers, confidence men, crusading clergymen, surveyors, railway builders, miners, merchants, entertainers, horse thieves, and bootleggers all travelled through, lived, and died in the Pass.

The large number of single men brought to the Pass by the construction of the Canadian Pacific Railway in 1897, and later by the development and operation of the mines, came with the scourge of alcohol abuse and its companion: crime. Bootleggers, who began stills in shacks in the bush during the railway construction, continued their enterprise even as the number of hotels and bars increased. In 1910, the Frank newspaper bemoaned the common drunkenness of miners, particularly after payday, when they spent so much of their earnings on liquor that many were left without money to pay their bills. At closing time on paydays, Mounties in red serge would ride horses down the main street of the town "and fire their revolvers into the air to break up the mob that poured from the saloons."[4]

In 1909, Corporal Fred Moses of the Royal Northwest Mounted Police (RNWMP) detachment in Frank listed some of the infractions of the law that he had enforced, as well as the respective fines that the miscreants were ordered to pay: "the sale of liquor without a license, $50; drunk and disorderly conduct, $25; keeping a disorderly house or a gaming house, $100; the theft of a ride on the CPR and possession of a loaded revolver, $20.90; commission of an assault upon his wife, $5; drunk and disorderly conduct, $2; possession of a revolver on his person when arrested, $20; frequenting a house of ill-fame, $25."[5]

Prostitution constantly aggravated the morals of the clergy in the towns of the Pass, but their public protests were to no effect. Men fought and whored despite the best efforts of authorities both moral and legal. What was described as a great stone and bottle battle erupted in the area of Coleman known as Bushtown in 1908. Bank employees were given target practice with handguns, and on at least one occasion in Lille, an employee had to fire his weapon into the floor to interrupt an attempted robbery. Newspapers complained of too little policing and incompetent prosecutors. Single, young immigrant men, rootless, without the commitment to and responsibility for a family or community, were often the problem.

Sports such as hockey, football, and baseball were popular among the miners and absorbed some of their energy. Many men became members of lodges that acted as surrogate families. The lodges offered social opportunities in the form of lodge dinners, smokers, and mutual support, but also made funeral arrangements for members.

Up until the end of the First World War, ethnic and religious barriers prevented social interaction. Immigrants tended to stick together for support. In an English-speaking country, those who could not speak English well, or at all, were marginalized and relegated to the poorest-paying jobs. The inability to understand instructions from an employer, or to understand or communicate information that had an immediate bearing on safety in coal mines, made the non-English-speaking immigrants less valuable to management than native English speakers. The result was that the Canadians, British, and Americans tended to become the managers and bosses. The fact that immigrants associated only with members of their own ethnicity also meant that English became more difficult for adults to learn. Their children, however, would take to English much more easily, as a result of their Canadian educations. For some immigrants, the decision to move to the Crowsnest Pass instead of large metropolitan centres such as Winnipeg, where more English could be learned and better jobs secured, became a source of regret. The barriers for their children, on the other hand, would eventually vanish, because they were educated in the same schools and spoke the same language as their native English-speaking counterparts.

Thus, ethnic, religious, and social support organizations played an important role in the miners' lives. In Lille in 1906, seventy-two Italian labourers formed the Italian Benevolent Society to help fellow countrymen in need, sickness, or death.[6] Other lodges were formed in Coleman and Hillcrest. A group of Ukrainians also built a hall in Hillcrest that was run by a socialist group, the Ukrainian Social Democratic Party (USDP). When Communists took over Ukraine, however, the USDP could no longer muster the support of Ukrainians to run the hall. By 1918, there was no longer any organized Ukrainian community in Hillcrest.

Lodges such as the Masons, Oddfellows, Knights of Pythias, and Eagles offered a sense of belonging and camaraderie that was generally missing among single men. More importantly, the lodges played a large charitable role in the communities, providing financial assistance to members, and to those in the general community. The absence of a welfare system resulted in a strong sense of community and recognition that common problems could be solved by common voluntary effort. Involvement in community became essential to survival. Furthermore, miners were known for their generosity. If a mining family were to lose its house to a fire, for example, other miners would gather and help build a new house, although such aid seldom crossed ethnic lines.

Children were generally expected to contribute to the welfare of their families, but as with youngsters everywhere, practical jokes became a way to beat the boredom. Frank Simister grew up in the Crowsnest Pass just after the turn of the century. As Simister tells it, the town of Michel at the west end of the Pass paid the Michel Hotel to prepare meals for prisoners in the jail. As a boy, Simister was paid twenty-five cents to take the meals from the hotel to the jail. Not one to miss an opportunity for a prank, he once locked the policeman in charge of the jail, Fred Lars, inside for three days.

Simister recounts the story of a man named Pat Groman, who used to go to Elk Valley packing horses for prospectors or hunters: "He left 14 head of pack horses near the Hotel. We cut the buggers loose, and scared them all over hell. Then we got two dollars to go out and catch the horses."[7]

Until opera houses were built in many of the Crowsnest communities, entertainment was hard to come by. Simister describes how hungry the men were for culture. "Around 1906, there was one hotel," he recounted. "Tom Crane had it. He used to have a great big gramophone out on the veranda, and they used to go out and listen to it, the only gramophone around in those days."

By 1909, an entrepreneur had begun showing movies in Fernie. By 1912 they were screened nightly, and by 1914 Fernie had three movie theatres. That same year, Joe Fumigalli bought two movie projectors and opened a movie theatre in the Hillcrest union hall. Movies by filmmakers like Charlie Chaplin were popular among patrons, who paid 25 cents for an adult admission and 10 cents for a child. Because the movies were silent, their messages and entertainment value crossed ethnic lines. Music had a similar effect. In 1914, Fumigalli sold one of his projectors to Billy Cole in Bellevue and helped him set up his own theatre in the old Bellevue union hall.

Music also became a popular recreation, and many miners and their children learned to play musical instruments. Robert, James, and Alex Petrie performed with the Frank Methodist church orchestra. Hillcrest boasted Alberta's first symphony orchestra, organized in the mid-1920s by Swiss immigrant Walter Moser. It consisted of talented amateur musicians from the nearby communities and hired itself out for performances, concerts, and dances. Bellevue and Coleman both boasted miners' bands, which played at funerals, parades, and other functions. Entertainment of another kind preoccupied some of the single men. Most people in the Pass took prostitution for granted. As Simister tells it, "There used to be three whorehouses down in Natal, just past Natal. Well, I used to drive men down there. I used to get a dollar to take my loads down. Big Helen in the white house, she used to give me a dollar if I took customers to her place first."

Many of the families who lived in the mine camps of the Pass, in particular the Eastern Europeans, kept livestock in their yards, including cows, hogs, and various kinds of fowl, to supplement their income from the mine. Most families grew vegetable gardens, which was far cheaper than buying vegetables from the grocery store. Chores, as well as school, kept children busy. They would round up the cows, milk them, and often feed and water the livestock before and after classes. Martha and Frank Labonne, children of French immigrants, fetched their cows to milk them, and delivered the milk in lard cans to their customers in Hillcrest to earn extra money for their family. Their father Francois worked in the mine with his two brothers-in-law, Emile and Leonce Chabillon, as well as a cousin, Alfred Salva. They were typical of many mining families in the Pass, connected by marriage and often by language and ethnicity.

As children in the latter half of the 1920s and into the '30s, Vasyl (Bill) and Mike Frolick, children of George and Mary Frolick, would deliver milk from their two cows to customers. Chickens supplied the family with eggs and meat, and the cows provided milk, butter, and cheese, as well as manure for their garden.

In general, women did not earn wages. They kept house and raised the children, cooked, cleaned, and washed clothes. Some worked as schoolteachers, others as domestic help, or seamstresses.

With Canada on the gold standard, inflation was not the problem it would later become. In 1914, one dollar had the same purchasing power as $18.34 in 2007. The Canadian dollar lost more than 94 per cent of its value between 1914 and 2005. In 1914, a loaf of bread cost five cents, versus $1.79 in 2005. A pound of butter that cost $0.25 in 1914 cost $4.01 in 2005. In the Crowsnest Pass in 1914, a shopper could buy two pounds of coffee for $0.85, a pound of clover honey for $0.30, and a sliced, cooked ham for $0.40 per pound. A woman could buy a wool dress for $6.50, and a man a made-to-order suit from $18. Tennis shoes cost $4.70 a pair, and gym shoes $2 a pair. For entertainment, an adult could see a silent movie for twenty cents, or a dime for a Saturday matinee.

The new century brought more than just the railway to the Crowsnest Pass — it also brought telephones. The Frank *Sentinel* reported in January 1902 that "[t]he telephone line being put in between Frank and Blairmore by the Alberta Mercantile Company is about finished, the wires being now well within the town limits."[8] In June, four men incorporated the Alberta Telephone Company: a store operator in Blairmore, Henry Lyon; a merchant with a store in Frank, Duncan McIntyre; a Dominion Land Surveyor, Joseph E. Woods; and the first settler on the Frank town site, Joseph Montalbeti. Despite the destruction in 1903 of part of Frank by the slide on Turtle Mountain, the project continued, and the entrepreneurs extended the line between Lyon's and McIntyre's stores "east to Burmis, and west to Coleman with connections to the mines at

Hillcrest, Bellevue, Lille and Passburg...."[9] Lodgepole pine trees carried the lines. In 1908, lines connected Calgary and Fort MacLeod with the Crowsnest Pass. By 1909, more lines connected towns and mines at the east end of the Pass in British Columbia.

In 1910, workmen completed a road that made it possible to drive a motor car from Fernie to Lethbridge, although not easily, as the road was rough, muddy when wet, and in some places little more than a rutted wagon trail. Two years later the road underwent extensive improvements.

Politically, many Hillcrest miners entertained socialist and Marxist beliefs. Coal miners in 1914 were, in general, not well educated or well read, which meant they were vulnerable to the demagoguery of Marxist union leaders. Socialism, or "progressivism" as it was known at the time, had become a strong political force in Alberta, and continued to drive the ideology of miners long after Alberta voters had turned away from its anti-individual, anti-free enterprise message. Among miners, these beliefs were held passionately, and those who rejected socialism generally kept their thoughts to themselves.

Frank Sherman, the founder and president of District 18 of the United Mine Workers of America (UMWA) from 1903 until his death in 1909, led radical unionism in the Crowsnest Pass. Born in 1869 in Gloucester, England, Sherman had worked the coal mines of South Wales, as had other union leaders in the Pass. He founded the mouthpiece of the union, *The Fernie Ledger,* and edited the paper from 1905–1907. From 1907–1915, the union turned the paper, also known as *The District Ledger,* into one of the leading organs of socialist thought in Canada. In 1908 Sherman ran for the Socialist Party in Calgary on one plank: the abolition of capitalism.[10]

At the 1909 UMWA convention, delegates also called for the expropriation of what they called "coal barons," and demanded public ownership and "democratic control" of collieries. The resolution's wording was identical to that of a resolution adopted by the international convention a few weeks earlier. This stream of political thought adopted by the UMWA contributed to an antagonistic and troubled relationship between miners and mine operators. When a miner believed what the union told him, that he and other miners "owned" or ought to own the coal mines by right, he would naturally feel wronged by and antagonistic toward the mine owners and operators who employed him, paid him (what he considered to be too little), and told him what to do in the mine. Never mind, for instance, that the plethora of unskilled labour in Canada resulted from the massive immigration of unskilled labourers, that labour itself was a commodity, and that its value dropped as it became commonplace. Miners became outraged when the drop in the value of their labour manifested itself as a drop in pay. They knew that a "closed shop" would give them the leverage to gain the wage increases they wanted, whatever the effects might be upon individual mines or the costs of goods and services reliant upon coal for transportation and heating.

But just as the price of labour dropped when labourers were abundant, when industry faced a shortage of labour, it had to pay more for it. During the First World War, when so many young men had left to fight the war overseas, very few men were left in Canada to mine coal. This meant that during the later war years, coal miners could earn more per day than any other workers in Canada.

Before the war, immigrants had poured into Canada, and the leadership of District 18 was acutely aware of the problem this created. At their eighth annual convention in Lethbridge in February 1911, the district's vice-president told members that the union had been active in pressing the federal government for immigration restraint, while at the same time had not argued for a complete stop. Rather, the union leadership objected to what the vice-president called "misrepresentation of conditions as effecting the workers of this country to the workers of Great Britain."[11] The alleged misrepresentation by advertisements trumpeting the benefits of immigration to Canada led many Brits to scramble for passage to a land they believed to offer abundant work. Those who came faced the often-harsh reality of too many men chasing too few jobs.

Politics is seldom simple, of course, and in Alberta and among Crowsnest Pass miners, this was as true as ever. Between 1905 and 1913, five prominent leaders of District 18 of the UMWA ran for office in Alberta general elections — not for the Socialist Party, but for the Liberal Party, a populist party that claimed to defend the rights of workers, farmers, and immigrants. The old-line politicians usually stood for free land, free trade, railway competition, Oriental exclusionism, and other ideas with which the coal miners agreed. Liberal MLA Charles Cross of Edson quoted Marx in the Alberta Legislature and recommended him to his colleagues. Cross was no eccentric backbencher. In fact, he was the attorney general under Alberta's Liberal regime.

When Lethbridge Mayor William D.L. Hardie, on the invitation of the UMWA, addressed the delegates on the first day of their 12th annual convention, he stated that although he was not a socialist, he was "very socialistically inclined. I am not one to get on a platform and advocate extremes in any particular thing. I believe all social advancement has got to be progressive, but we do not need to go to extremes although we desire to be modern." Hardie's statement equates modernism with progressivism, or socialism, and although Hardie spoke only for himself, it illustrates the degree to which progressive politics had taken over the province. The fact that the union appointed a press committee to censor all press reports coming out of the District 18 convention may have emboldened Hardie to speak his mind away from the public view and scrutiny. What he seemed to imply was that socialist policies were the wave of the future, but that a *slow* march "forward" was preferable.

His speech was, in part, a plea for union members to support the war, something District 18 was not inclined to do. *After* the war, said Hardie, they should get together to deliberate upon principles for "our social advancement." He also urged members to pursue wise, moderate, and careful demands for wage increases in such hard times. Yet wages throughout the war would increase for miners due to a scarcity of labour, and not because the miners pushed for raises.[12]

Sherman sometimes acted and spoke contrary to what some might expect of a Marxist. In 1904 he strongly implied that a producers' cartel of some kind might be useful to organize against what he saw as the conspiracies of the monopolistic Canadian Pacific Railway, long held in contempt by Crowsnest miners. His proposal remains as vague as his accusation against the CPR, although his language suggests he felt that coal mine operators should act together, in effect as a labour union, to force the CPR to raise the price it paid for coal by withholding product. During the Lethbridge coal strike of 1906, after drunken, militant union men engaged in pitched battles with the RNWMP, which led to arrests, followed by efforts by the crowds to free the detainees, the authorities declared martial law. Sherman then counselled against all demonstrations. He also ardently supported temperance, something miners were not likely to support. Sherman, a radical unionist and a complex man, entertained ideas riddled with contradictions. It takes only a small leap of faith to conclude that his Marxist influence energized unnecessary conflict in the Crowsnest Pass coalfields.

One might think that Sherman's opinions, suggestions, and recommendations would be anathema to the government, but apparently not. On March 16, 1909, Sherman wrote a letter to Alberta's Minister of Public Works, William H. Cushing, urging Cushing to hire Frank Aspinall of Edmonton as a mines inspector. Sherman described Aspinall as a thoroughly competent man who had worked himself up through the ranks. Sherman felt obligated to add that he sincerely trusted that Cushing would be re-elected, "even though you know I am opposed to your side in politics." At the time, Liberals controlled the Alberta Legislature. Certainly, in his writing of the letter, Sherman presupposed that he had a certain amount of influence with the provincial government. In fact, the province did hire Aspinall as a provincial mines inspector, but whether Sherman's letter influenced the hire remains an open question.[13]

Sherman also was directly responsible for the insertion of an anti-Asian clause in the international UMWA constitution, although he reversed his and the local's position in 1909. He had helped open the Pandora's box, however, and his reversal would be ignored. His successor at the 1909 district convention, a member of the Socialist Party, revoked it and mocked the "Slant-Eyeds." The contract signed to end the 1911 strike, however, contained a clause that referred specifically to Chinese labour. It said that the union did not prohibit the employment of Chinese workers in or around the mine, but they had to be paid the prevailing rates. The clause concluded, "Present conditions to prevail."

At the time of the Hillcrest disaster in 1914, not a single non-European worked for Hillcrest Collieries. In B.C., anti-Chinese sentiment among miners had been strong since the Nanaimo explosion in 1887, which had killed as many as 150 men and was blamed on the Chinese. When seventy-five men died the following year in an explosion at Wellington, again the Chinese were blamed, and European miners in B.C. pressed for a ban on Chinese workers in underground sites. The miners argued that the Chinese, who could neither speak nor read English, threatened the mines' safety. Many of the European miners, of course, were also unable to speak English, and many of those who could were illiterate. Chinese miners were known to work hard for less money than their European counterparts, which may have been the real objection. At the end of May 1916, a District 18 member moved that union resident officers be instructed to convey a veiled threat to the Western Coal Operators Association: "that the District will not feel responsible for any action that may follow the non-successful exclusion of Yellow Labor." The motion was carried.

Oriental labour, of course, was not the miners' only grievance. In general they were hostile to capitalism, mine owners, the military, police in general, and the RNWMP in particular, whom the miners saw as siding with the mine owners. On the other hand, in the face of violence, or threats of violence, the RNWMP were obligated to respond as they would to any lawbreakers. Although there is no evidence that the Hillcrest local was a target, one tactic sometimes deemed necessary by police to keep tabs on threats of violence by radical union members was the infiltration of unions by undercover police officers. Cortlandt Starnes used undercover operatives to monitor criminal activity while he policed in the Yukon during the gold rush. The problem, from a policing point of view, was that the Mounties were vastly outnumbered by miners who had easy access to dynamite and the tremendous destructive power it represented.[14]

The suggestion that miners might use dynamite to try to settle grievances was posited during a strike at Lille in 1907. Inspector T.S. Belcher blamed the strike largely on "foreigners," and reported that "these foreigners seem inclined to make trouble." Belcher recommended laying charges against the miners for the illegal strike based on complaints by the mine management, but sought advice before he did so, and commented in his report to Superintendent Primrose that "probably if a few of them were fined for going on strike it might have a good effect, or it might make further trouble." Primrose in turn sent a telegram to the commissioner in Regina. "This place Lille," he reported, "is very much of a mountain fastness, being bad to get at. The last time any friction took place at Lille the thoughts of those people ran to the using of dynamite, by throwing or otherwise." Primrose concluded his report with a reminder that he had pointed out a few days ago that he was short of men, and that "if we put our hands to the plough to prosecute these people, we want to be able to go through with it."[15] In other words, if the Mounties decided to enforce the law, they would need enough policemen at Lille to deal with any possible concerted actions, including violence, by the striking miners. Marxism brought to the miners a belief that the forces of law and order were against them, a belief that arose out of a sense of perpetual grievance, and of frustrated entitlement — i.e., that they were entitled to certain things that were unjustly denied them by mine owners, management, and the law itself.

The appearance of a miner at Hillcrest Collieries in 1914 was both distinctive and commonplace. It was commonplace because all miners of the period dressed similarly, and distinctive because their clothing and paraphernalia instantly identified them as miners. A miner walked the fine granular coal floor of the mine in hobnail boots. His clothing consisted generally of a long-sleeved shirt, a rough jacket or coat, and loose-fitting cotton trousers. For warmth, he layered clothing that could be easily peeled off or added, for the sulfur-smelling air of the mine blew cool in summer and moderate in winter. He generally wore a cloth cap of some type; only the mine horses wore helmets of leather. On his cap or jacket he pinned his brass mine check. The numbered checks allowed management to account for who was in the mine, where they were working, and the lamps handed out at the start of a shift, as well as to identify a body in the event of a catastrophe. Every individual who entered the mine, including inspectors, management, and visitors, was given a check to take with him into the mine — usually a round brass numbered tag that was pinned to his clothing. No man ever entered a mine without one, nor without a Wolf safety lamp, which gave off a dim light (less than that of a candle). Slung over his shoulder hung a lunch pail, usually stacked with compartments for drinks and sandwiches. Contract miners would carry wood-handled steel picks to pry out the coal and chip off rock. Others carried long hand drills to penetrate the coal at the face with holes for the fire boss to insert

Monobel for blasting. The miner began his shift rested; he ended it physically exhausted. He walked into the mine clean; he left covered in a layer of coal dust so fine that it entered every crevice of his body, and resisted the most furious assault by water and soap. A man could often be identified as a coal miner by a cursory examination of the skin of his hands, which displayed very fine blue colourings. When a miner wielded a pick, the hard coal flew up and penetrated his skin. The small pieces of coal left a slight blue colouring under the skin, in effect a tattoo that could never be removed or washed away.

At Hillcrest, virtually everyone in the town faced constant reminders of the source of their incomes. Coal dust generated by the activity at the tipple often drifted widely over the town, depending upon the direction of the wind, and settled on everything, including the clean laundry hung out on clotheslines by the women of the town.[16] Sometimes they were forced to wash clothes two or three times. While coal dust was a nuisance in Hillcrest, it was also a fact of life, like snowfall in winter and rain in summer.

Perhaps one of the most important aspects of miners' lives in Hillcrest was that, unlike some company towns in the Crowsnest Pass, miners in Hillcrest could, if they wished, own their own homes, and many did. Miner George Frolick owned a house in Hillcrest, and his son remembered Hillcrest Collieries as a benevolent company that supplied running water and electricity without cost, and, he suspected, provided a good portion of the budget for the school, which was more elaborate than a typical one-room country school.[17] Hillcrest's school boasted separate rooms and a teacher for each grade. This ability to purchase and own their own homes gave the lives of Hillcrest miners stability absent from the lives of miners at other mining camps, and a sense of investment in the community. Life revolved around home, church, the union hall, and, above all, work.

CHAPTER 3
C.P. HILL: AN AMERICAN
IN THE COAL FIELDS

The Hillcrest Coal and Coke Company existed because of the enterprise of one man, born American, Canadian by choice. Charles Plummer Hill entered his life in the small town of Seaford, Delaware, in 1862, during the American Civil War, under the administration of Abraham Lincoln, and before the death of George Armstrong Custer in the grassy hills of Montana. His family was not wealthy. As a boy, Hill attended public school, and began his working life as a mechanic with the Jackson & Sharp Manufacturing Company of Wilmington, Delaware. He seems to have developed an interest in railways early on, and held various positions in the Department of Construction, Designing, and Operating of the Pullman Palace Car Company of New York and with the Northern Pacific and Great Northern Railways in St. Paul. He discovered, though, that the life of a mechanic and engineer was not the life he wanted; instead, he hungered for adventure.

The lure of the great northwest beckoned Hill, and like many young men of his day, he decided to head west to make a name for himself in territory that was largely undeveloped and unspoiled. He found a job as a U.S. deputy collector of customs in 1892, which made him the inspector of 800 miles of boundary line between British Columbia, Alberta, Montana, and Idaho. For a base of operations, he created a small border post just south of Creston, B.C. that was named Porthill in his honour. The job proved ideal for the restless, ambitious Hill, because it allowed him to earn a living while he explored the area on horseback — heady stuff for a city boy from the Eastern Seaboard. He travelled wherever a horse could walk or be led, slept in a tent or under the stars, and lived partly off whatever game the land provided while he searched and prospected. Hill became toughened by the land and the life, but eventually opportunity beckoned, and based on what he had discovered in the course of his exploration and prospecting trips, he decided to make his move. He quit the U.S. Civil Service in 1899 and moved to the Kootenays, from where he began to prospect full-time. His work quickly paid off. Hill explored and undertook development work in the Kitchener Iron Range, which he later co-owned with the CPR. During his explorations, Hill discovered large deposits of coal in the Crowsnest Pass.

In later years, Hill claimed that he spent the year 1900 in the Klondike, where the gold rush had drawn thousands of adventurers who hoped to strike it rich in the Yukon goldfields.[1] If Hill was in the Klondike in 1900, he was not there for the full year. Perhaps he saw that the likelihood of striking it rich in the goldfields by that time was so unlikely, and the effort such a waste of time, that he returned to southern Canada, convinced that his way to success lay through mine development — not of gold, but coal. The work of coal mine development would certainly be more difficult, and he would not get rich quickly, but for that reason he would face less competition. Hill arrived in Alberta sometime in 1900, where he proved himself to be an able prospector for coal, and his work quickly proved fruitful. Hill's exploration work led to the development of several mines in the Pass, and his decision to avoid the Klondike proved wise.

With the construction of the Crowsnest Pass Railway, Hill became acutely aware of the growing demand for coal, and was well aware of its abundance in the Pass through his own exploration and prospecting. Massive blankets of coal, the remains of a more than 150-million-year-old forest, underlie much of the Crowsnest Pass, as the Upper Jurassic–Lower Cretaceous beds of the Mesozoic lie sandwiched between layers of the Kootenay sandstone. Hill understood that the key to success lay in the quality of the coal. Not only did the CPR carry large shipments of coal and coke from mines in the Pass to smelters, but it also burned coal in its steam locomotives. For steam coal, railways demanded coal that was low in ash byproduct, free of impurities, and hot-burning. The better the quality, the more efficient the locomotive. Anyone who could supply high-quality, low-cost coal would position himself right in the path of the CPR's steam requirements, and could occupy a profitable, if single-market niche.

Thus, in the Crowsnest Pass at the turn of the century, Hill became one of the major players in the exploration for coal. In 1900, with financial support from three Montana men of influence, U.S. Senator Thomas H. Carter, H.L. Frank of Butte, and W.S. Gebo of Red Lodge, Hill sank a shaft to a depth of 433 feet on what became known as the Burmiss Property. Frank, Gebo, and another partner, Henry S. Pelletier, later formed the Canadian American Company and began development of the coal seams at Frank. Mark Drumm, who became editor of the Frank weekly newspaper, began his career in the Pass as secretary-treasurer of the Canadian American Company. It was for Mark Drumm that Drumm Creek was named, and which became the source of water for the Hillcrest mine and town.

During this period, Hill also ran the first tunnels at McLaren's Lumber Camp on York Creek on the seams that were later owned by the International Coal and Coke Company of Washington State.

Later, after he became a mine owner, Hill was seen by union men as part of the privileged class, the enemy. But Hill had earned his position and wealth. At this stage of his career, without money or financial backing, and with little but determination and a willingness to work hard, he dug and sweated through the first 200 feet of the prospect himself with pick and shovel. He became accustomed to manual labour, and knew what men were capable of doing physically. By 1901, Hill resided in rented lodgings in Blairmore, and described himself as a mining promoter. That same year, he ran a cross-cut tunnel on the Blairmore property, and with Mr. Wilson of the Crow's Nest Pass Coal Company, explored the Holloway property that was later developed into producing mines. Hill also prospected on the south fork of the Crowsnest River, where he discovered good coal and purchased and prospected the Wolstenholme property. But of all the coal that Hill discovered, he saved the best for his own mine.

On July 16, 1902, in the Crowsnest Pass, Hill staked the east half of section 19 after he had purchased the rights from a group of settlers. On this land, he would build Hillcrest, both the town site and the mine. He bought additional land nearby, and began to promote and develop the property. In 1904, Hill became a Canadian citizen, and married the daughter of a Charlottetown man who had also come west. Hill's marriage to Enid Mary McLean would last until her death in Victoria in 1932.

Shareholders in the Hill Crest Coal and Coke Company included members of Hill's extended family in Hillcrest, his father-in-law, and a brother, as well as one of the people from whom Hill had purchased the rights to the property, George Mills of Frank.

Hill staked out the town site on a park-like tract of land at the east foot of Turtle Mountain, about 2.5 miles east of Frank. To some extent, Turtle Mountain protects the town from the wind, which could and often does blow with great intensity through the pass. Drumm Creek provided an abundant supply of clean water, both for the mine operations and for the town, and the town was closely surrounded by four other coal mines at Frank, Bellevue, and Passburg.

Hill began the development of his mine, however, with more ambition and determination than capital. One of his earliest employees, John Kunesky, arrived in Canada from Ukraine in 1904. But finding only low-paying work in Winnipeg, he left for better pay in Hillcrest.[2] Hill, still just a prospector at the time, hired Kunesky and five other Ukrainian men to build a road through the bush to the future site of the coal mine. He agreed to pay them $2.25 per day, as well as $3.00 per day to the miners who opened up the mine. When the cheques bounced, Hill revealed that he was actually unable to pay anyone, but promised that he would when the mine opened. With no money to go anywhere else, the men agreed, with sinking hearts, to work anyway. The promise of food and a roof over their heads spoke loudly. Hill put the men up in shacks that he had built and provided them with food while he secured investment capital from Eastern Canada. On January 31, 1905, Hill incorporated the Hill Crest Coal and Coke Company. His investors included a diverse group of individuals: Ontario members of parliament, a Quebec MP, a CPR land commissioner, a Winnipeg bank manager, an American railway contractor, a Toronto physician, an insurance agent from New Brunswick, and others. Six months after he had first hired the men, Hill paid them the money he owed them. Hill Crest Collieries was in business, and Hill's work was about to pay off.

Among his first difficulties was the fact that Hill's coal was buried some distance from the CPR tracks. To get the coal to the CPR line for shipment, he had to build his own short line railway over a distance of about two miles. This made Hill's coal more expensive to mine than coal from other, closer mines in the Pass. By November 3, 1905, his men had driven tunnels into two coal seams to prepare for shipment, but were forced to wait for another crew to finish the railway from the mine to the CPR tracks. Without a rail connection, Hill's coal could not be sold or shipped in any profitable quantity. In December, Hill's aboveground workforce totaled ninety men. Only twenty-four laboured underground. By late March that number had dropped to sixteen. It wasn't until March 22, 1906, with the completion of the short line, that the mine could ship coal. The loaded cars were assembled in a yard, picked up by the small steam locomotive called a dinky, and hauled to and dumped at a log tipple several hundred yards away. A conveyor delivered the coal to a loading bin alongside standard gauge tracks about 125 feet below. A company train crew then pulled the loaded cars and returned empties between the tipple and the CPR sidings, up and down the hill from the tipple and across the Crowsnest River, on a curved wooden trestle.

By June 7, with shipments underway, the number of men underground had leaped back to forty in two shifts. Hill's men produced a modest 150 tons per day from the no. 1 seam. However, in Hill's haste to ship coal and start earning money in the face of mounting expenses, he chose perhaps the oldest and crudest method around of ventilating the mine: a furnace placed at the bottom of the upcast. Air heated by the furnace would rise to the surface and draw air from the mine to replace what was expelled. It was crude but effective. It was also dangerous, because with the air came the danger of firedamp. The peril did not go unnoticed by the provincial mines inspector for the district, Elijah Heathcote, who reported his observation to the Alberta deputy minister of public works, John Stocks.

Stocks had been appointed in 1905, when the province of Alberta was formed. Born and educated in Quebec, he worked for the CPR for twenty years, until 1901. He began as foreman of construction, but moved up rapidly through the ranks to superintendent of construction and maintenance, a position he held for eighteen years over a district that extended from Swift Current west to Laggan (Lake Louise). Stocks was largely responsible for the system of roads in Alberta after 1905, as well as the plan of trunk roads connecting all the important points of settlement, with branch roads extending outwards.[3] He cultivated an intimate familiarity with the CPR and its needs, particularly the importance of high-quality steam coal.

In a sternly worded letter to Hill Crest Collieries, Stocks drew attention to the fact that the last inspection report on the Hillcrest mine was anything but favourable, and ordered the general manager to discontinue the use of the furnace for ventilation at once, and immediately begin ventilating the mine with a fan. The deputy minister also ordered that natural ventilation be employed in the mine whenever possible. He left the number of men to be permitted to work in the mine to Heathcote's discretion until a fan could be installed.

Heathcote reported on September 19, 1906 that the use of the furnace had been discontinued, the number of men underground reduced to a maximum of thirty-seven, and the only ventilation in use until Hill's steam-powered fan arrived "natural"—i.e., none at all. Years later, an outside expert employed by the CPR would observe that the natural ventilation at the Hill Crest mine was excellent. In 1906, so early in the mine's life, firedamp posed less of a problem than it would later as the mine penetrated deeper under the rock. The greater the overburden, the more firedamp released. Heathcote also noted in his report that the mine had closed for the time being due to labour and financial difficulties.

With the arrival of 1907, Hill embarked upon a serious effort to further promote the property in Eastern Canada. That spring, a man lost his life at Hill Crest Coal and Coke, but this first fatality did not occur inside the mine. The engineer was killed when the Hillcrest locomotive jumped the tracks and fell from a trestle. It was an inauspicious start to the year. In the late spring, miners in B.C. and Alberta went on strike, forcing the suspension of mine operations. Operators had formed the Western Coal Mine Operators Association in late 1906 to unify and strengthen the bargaining position of operators in contract talks with the union. By April 1907, the Association consisted of Pacific Coal of Bankhead, the H.W. McNeill company of Canmore, Canadian American Coal and Coke of Frank, the Breckenridge and Lund Coal Company of Lundbreck, Western Canadian Collieries of Blairmore, International Coal and Coke Company of Coleman, Maple Leaf Coal Company of Bellevue and Lille, and the Crow's Nest Coal Company of Fernie and Michel. Hill Crest Coal and Coke did not join until January 25, 1909.

Shortly after the operators had formed their bargaining group, the federal government passed the Industrial Disputes Investigation Act, also known as the Lemieux Act, to prevent strikes and their consequent damage to the economy, to companies, and to individuals. The Act forced unions, before striking, to take their grievances to conciliation boards appointed by the government. The union became unhappy with this procedure. By the end of April 1907, most of the district had gone on strike despite the Act, as the board met to discuss the owners' and miners' differences. When Hill returned from the east and was faced with the summer strike, he made use of the time to continue with mine development work. First, he drew up plans for a town site, then published a brochure to promote the mine and town. The brochure included a list of lots that he would put up for sale. The most expensive lots on the corners of Main Street or Broadway would sell for $400. Prices varied down to $125 on the corners of White Street, or $100 on the inside of the White Street block. The strike ended on July 24, and production resumed.

By mid-November, with the new fan in operation, Inspector Heathcote reported that it produced 11,200 cubic feet of air per minute in the main gangway and 20,250 cubic feet per minute for the slope workings, and that the production of the mine had leaped to 500 tons per day. He also reported one other significant fact: open lights were used throughout the mine for illumination. These were cap lamps, or small oil lamps attached to the front of miners' cloth caps. They were extremely dangerous in the presence of firedamp.

On February 7, 1908, Heathcote wrote in a report to the deputy minister that open lights were still being used at Hill Crest, but that he had spoken to Hill about the installation of safety lamps, which were now mandated by law. As well, gas was reported in the mine on January 21. Hill ordered the safety lamps.

Heathcote also wrote that Hill planned to install an air compressor plant so that coal could be handled with air motors instead of horses. The air motors — large tanks of compressed air that drove the wheels — required a compressor just as a tire in an automobile requires one to fill it with compressed air. The compressed air then drove the haulage motor in safely, without the threat from flames in a steam engine, or hot exhaust from an internal combustion engine. The Hill Crest mine used horses to haul full trips from the chutes on the levels to the slant where a wire rope attached to a steam-powered winch would haul the trips to the surface. Heathcote recommended that Hill be granted permission to use open lights for 12 months until the compressor plant was installed, but only along the level for 1,800 feet from the entrance.

Then, on February 10, just a few days after Heathcote's report, the inevitable happened in the face of open lights: a gas explosion, ignited by their lamps, severely burned two Hillcrest miners. The fire boss, John C. Clark, was supposed to examine the mine thoroughly, but did not. The subsequent investigation determined that the two men burned did not see Clark, and since no warnings were posted, they had assumed that all was clear. Inside the mine, one of the miners had climbed to the top of the chute and set off the gas with his open lamp light. The subsequent explosion burned both him and his colleague. Clark's career at Hillcrest was over.

Hill no doubt felt himself caught between the demands of the miners (who opposed safety lamps), their union, the provincial inspector, and his own personnel problems. Exasperated, he failed to report the accident, as required by law, knowing full well what the result would be: an immediate demand by provincial authorities for safety lamps to be used exclusively at Hill Crest.

This did not, however, sit well with the miners — as Hill knew it wouldn't. On April 6, the local wrote a letter to John Stocks, in which they asked him to instruct the inspector of mines "to be a little more lenient with us in respect to the safety lamps." The union argued that there was very little gas in the mine, and that the use of safety lamps "would be a considerable reduction to us under the existing agreement." In other words, the union argued that because of the enforced use of safety lamps, instead of the open lights preferred by the miners, they would not be able to dig as much coal. The letter prompted questions in the minds of Alberta government mines officials. Norman Fraser, of the Alberta Department of Public Works, travelled to Hillcrest and interviewed union officials. In private meetings, one admitted to him the danger of working the Hillcrest Mine with open lights. Two other officials admitted that they only wanted the safety lamps taken out of mines for financial reasons.[4]

The admission is revelatory. The letter from Fraser suggests very strongly that the UMWA wanted safety lamps removed not only from Hill Crest, but also from all other mines. Clearly, the issue was much bigger than the events at Hillcrest, and reveals that union officials, who knew well the danger of open lights, faced tremendous pressure from their own rank and file. It is a perfect example, not of greed on the part of miners, but of profound ignorance, and a consequent inability to recognize their own best interests. One feels in the entreaties of union officials a plea for mining officials to recognize the difficulty of the union's need to accommodate the will of the majority against the clear and present danger of open lights. This would seem to be the only explanation for their astonishing honesty in their admission of motives. They were clearly helpless and knew they would lose their battle — and, as expected, their request was rejected. On May 22, 1908, Heathcote reported to his superior that safety lamps were now used exclusively in the Hillcrest mine. It appeared to be the dawn of a new era in the Hill Crest Collieries.

Unfortunately, the battle over safety lamps was not yet over. On September 2, 1908, the lamp house burned to the ground in what was described as an "accidental" fire. Most of the safety lamps were destroyed. Desperate both to resume production and to accommodate the miners, Hill sent a telegram to Heathcote asking if he could work the mine with open lights again, but Heathcote refused. This was the beginning of a protracted battle between the two men. Perhaps not trusting Heathcote to represent the situation at Hillcrest fairly to Stocks, Hill wrote him an angry letter saying that "Mr. Heathcote should be taught that his duty is not to harass this Company any further." Hill explained that 200 safety lamps were destroyed in the lamp house fire. Bitterly, Hill noted that in the absence of safety lamps, Heathcote closed the mine. Hill defended his record on the protection of life and property as excellent and argued that all of the licensed men at the mine, as well as the miners themselves, agreed that the lamps were not necessary. Hill reiterated that even the union did not want them, and reminded Stocks that he got a letter from the union to that effect. Hill added that he believed in safety lamps as a preventative measure, but said Heathcote's closure of the mine was motivated by nothing but malice towards the company. Hill's problem was that as his mine sat idle, he was paying his employees to do nothing — a situation that would continue for as long as it took him to acquire new safety lamps. Contracts to supply coal also went unfulfilled, which would ruin the mine's reputation as a consistent supplier, and thus damage its future prospects. To further exacerbate Hill's cash crunch, he had continued with his expensive, albeit necessary development work. Electric lights had been fixed on the main gangway for a distance of 1,800 feet from the entrance of the mine, and an airshaft had been driven through to the surface at chute 28. To power the lights, Hill installed a generator at the power house. At the same time, Hill had to deal with personnel problems. On September 5, he appointed a new mine manager.

Hill's angry letter to Stocks had something to say as well about what, for Hill, was the exasperating issue of a wash house. Hill wrote that he had received a report from Heathcote strongly suggesting that the mine did not have a wash house, as required by law. Hill points out that the claim was false, as Heathcote had been in the wash house in Hill's presence. As for the criticism that the wash house was inadequate, in part because of its location above the boiler house, Hill reminded Stocks that it was placed there before the law was passed. One may infer that Hill believed his wash house should be grandfathered in light of the response he got from the union in the fall of the previous year, when he offered to build a wash house at a cost of $4,000 or more. He asked the union if the miners would patronize the lockers, through a small rental fee, in the same way miners did at other mines. In his letter, Hill says the response he got was that they would not rent lockers and did not want a wash house; he argues that the current wash house has hot and cold running water, and is located above the boiler house, "which is the proper place I consider except for the filth of these men whose only charge for a wash house made by this Company was that they should keep it clean. This is the only Mine that I know of that furnishes a wash house free to the men. At all other mines all lockers are rented." It was a statement of fact. Hill wrote that he had ordered 10-inch water pipe in connection with the new power house, and planned to build a new wash house when the line was completed, but not now.

While Hill fumed and his mine suffered financially, Heathcote wrote his own letter to Stocks on September 28. In it, he claimed that in the estimates of the Hill Crest Coal and Coke Company, Hill had reserved $8,000 to build a new wash house, but Hill wanted the men to pay a dollar a month each for the use of the lockers. According to Heathcote, the men told him in a meeting at the Imperial Hotel in Frank that they were not willing to take up lockers and pay to use them. In another swipe at Hill, he added that in a meeting later the same day at the Imperial Hotel, Hill was vague about how long it would be before a new wash house was built. He said that Hill told him he needed to lay a new pipeline and build a new power house before building a new wash house, but could not say how long this would take. Hill said it would be as soon as he could; in Heathcote's opinion, "this means anytime." Heathcote further explained further, somewhat contemptuously, that Hill's "scheme" was to drive a new rock tunnel to the coal seam and at this point install a new wash house and power house as soon as possible. Hill's "scheme" was made a practical reality in 1910 and 1911 by new owners who recognized the practical necessity of the plan. Heathcote says in his letter to Stocks that although Hill thinks Heathcote is harassing him unnecessarily, "I do not bear any malice towards Mr. Hill," adding that Hill told him that "he was going to try all in his power to get me fired, for not allowing him to use open lights in his mine, when his lamp house was burned up and also for a new wash house."

Two days later, Hill wrote another letter, this time not to John Stocks, but to the man at the top: the Minister of Public Works, William H. Cushing. Hill wrote about "what I consider many injustices imposed upon this company by your Mr. Heathcote, District Inspector of Mines." Hill claimed once again that there was no need for safety lamps, and that he had submitted to every request of the department, at great expense to the Company, which had never paid a dividend to shareholders. Writes Hill, "I am now tired of listening to the second hand complaints of our socialist mine committee as presented by your Mr. Heathcote." Hill charged that Heathcote was drunk in the Imperial Hotel in Frank on September 18, and said to the pit committee that he would compel Hill to settle the case as they wished it. Hill asked for an arbitration board consisting of licensed mine managers to rule upon the issue of the mine closure by the provincial inspector for lack of safety lamps, when safety lamps, according to Hill, are not required. Hill concluded that he refused to have his company's business again aired by a drunken mines inspector in a public barroom without a protest. The ministry had no choice but to investigate the accusation.

It appears that the Ministry of Public Works wanted to determine whether Hill's charges were accurate, but there may be more to it than that. The job of getting to the bottom of it was handed to Norman Fraser, who apparently took his investigation no further than reviewing the correspondence, memos, and other documents in the mine's file. In 1914, Fraser would serve as the mining expert for the UMWA at the Commission of Inquiry after the disaster. He would never have been hired by the union if he had been unsympathetic to union ideology and causes. Fraser summed up the dispute between Hill and Heathcote in a succinct six-page report, which he sent to Stocks on October 7, 1908.

The battle between Hill and Heathcote began, he pointed out in the report, in June 1906, when Heathcote complained in his inspection report that the Hillcrest mine had no certified manager in charge, as required by the Canada Mines Act. He reported Hill's dangerous use of a furnace for ventilation. Hill in turn complained that Heathcote agreed when he was at the mine that everything was fine, and then writing later that a great many things were wrong. In November, an inspection again found no qualified mine manager in charge. In March 1907, Heathcote reported that a certain report book mandated by the Canadian Mines Act was not kept at the mine. On March 25, Stocks told Hill that he had to use safety lamps if gas were found. In August, Heathcote criticized the ventilation and recommended the use of safety lamps, because he expected gas to show up when they got deeper under the mountain. In October, the ministry again pointed out that the manager did not hold a certificate. In November, Heathcote reported that open lights were still being used, and that gas was reported by the fire bosses on nine separate days. On February 7, 1908, Heathcote reported that open lights were still being used. Then, on the 10th, a gas explosion in the mine severely burned two miners. Hill had to promise to get safety lamps installed, and, at last, he did. In his summary, Fraser claimed that upon Heathcote's first visit to the mine, Hill became angry and used abusive language in the office, claiming that in all his time mining this was the first time he had been subject to inspection. As for Hill's continued claim that safety lamps were unnecessary, Fraser wrote that there was every reason to use safety lamps for accident prevention, as proved by the burns suffered by those two miners. He added that "a small local explosion of gas may at any time develop into a general explosion involving great loss of life." It would prove to be a prescient comment.

In Fraser's summary, he pointed out that although Hill claimed the company had not paid a dividend to shareholders, Hill wrote in his 1907 report to shareholders that the work at the mine showed good or better results than expected, and that the profit per operating account was $18,911.54. Fraser added, in response to Hill's request for arbitration by a board of licensed mine managers, that arbitration is not, according to the CMA, conducted by a board consisting only of licensed managers.

In his conclusion, Fraser pointed out that very little gas was reported in the Coleman mine before an explosion there that took three lives. At Lille, he says, management refused to introduce safety lamps because their mine contained very little gas, but almost immediately an explosion burned a man. In the end, Fraser's summary of the dispute between Hill and Heathcote defended the latter. In it, however, there is no mention that Fraser made any effort to contact Hill, nor of Heathcote's alleged drunkenness, his union sympathies, or his alleged unprofessional comments.

Hill was not a man who liked, or knew how, to mince his words, and his plainspoken nature made him eminently unpopular. The picture he seemed to present to others was of a recalcitrant, obstinate, abrasive, and consequently disliked man. In his defense, it can also be said that Hill clearly felt besieged on all sides. His single-minded purpose was to run a mine and create wealth. As the owner, however, he was obliged to appease various competing and hostile interests, including employees, mine inspectors, the union, and his investors. Together, they must have seemed to conspire against his personal success in favour of bankruptcy and humiliation.

We may assume, however, that no one wanted Hill's mine to fail. On the contrary, the CPR wanted it to succeed, with Hill or without him. They wanted the high-quality coal for their locomotives. The UMWA wanted it to succeed, but would have preferred that the union ran the mine. Its members wanted jobs, although many may not have understood what successful mine management demanded. The provincial government wanted it to succeed: it wanted jobs for men who would become trouble-some if unemployed, and wanted them to spend their paycheques in Alberta to boost the provincial economy. Hill's annual report seems likely to have been a puff job. He desperately needed to pay his stockholders a dividend to justify his need for more capital for needed improvements in what was clearly an undercapitalized mine. Hill's anxiety about the mine's future increased.

It should also be pointed out that conflicts between mine management and unions (with whom the provincial inspectors seemed to side) over the application and enforcement of mine regulations were common, and the hostility generated was con-siderable. As late as 1913, both the union local and the provincial chief inspector of mines both severely criticized Leitch Collieries at Passburg and Maple Leaf Coal Company at Bellevue for allegedly inadequate wash houses. In response to a complaint from miners that the wash house was unsanitary, the manager of the Bellevue mine pointed out to the provincial inspector of mines, John Stir-ling, that a new wash house was under construction, and that the old one was no different than it was when Stirling and his district inspector visited the site earlier and found no fault with it. He also pointed out that the management cleaned out the wash house twice a week, but the men themselves dumped water onto the floor instead of into the provided cesspool. The complaint echoed

Hill's comments from five years earlier, when he had complained that when the inspector was on site, he found nothing wrong, but later sent Hill a letter full of alleged infractions. The letter also echoes Hill's circumstances in its curt demands for a new wash house at a time when a new one was currently under construction.

Change rooms, showers, and locker rooms would in later years become taken for granted, but at the time, employers wondered how they had suddenly become responsible for the cleanliness of their employees. They also saw the miners' unwillingness to maintain the cleanliness of their provided facilities, or to pay a nominal fee to rent lockers, as ample proof that the miners simply did not want to assume responsibility for themselves. It seems likely that in the long run, mine management would have provided this convenience to miners anyway, as a means of attracting the best workers in a competitive environment. It is a telling reinforcement of Hill's position that after the long, bitter strike of 1911, a dispute that centred on whether the union should win a closed shop, the new agreement included a clause obligating the men at Hillcrest to pay $1 per man per month for use of the wash house.

The lack of objectivity that Crowsnest mine management believed they faced in the enforcement of mining regulations by provincial mines inspectors and the Department of Public Works is well illustrated in a letter dated February 14, 1913. It was written by the general manager of the Maple Leaf Coal Company at Bellevue, James Finlay, to the deputy minister of public works, John Stocks, in Edmonton. Before he became district inspector of mines for Crowsnest, Frank Aspinall had been the manager of the Maple Leaf mine, but Finlay had either let him go or fired him. Finlay, who claimed that he had managed mines in Nova Scotia, British Columbia, and Alberta without any negative reports, sent the letter in response to negative reports from Aspinall. Finlay blamed Aspinall for trying to find fault with the engineers and machinery department of Maple Leaf, and in particular with the fencing off of the machinery. Aspinall, he pointed out, had been manager when the hoists were put into position, but after Aspinall was dismissed, the mine management discovered that he had installed it incorrectly. As a result, they had to take out the hoist engine that Aspinall had set, make a new foundation, and replace it. Furthermore, he said, a boiler that had been installed under Aspinall's supervision had to be readjusted after Aspinall had left the mine's employ in order to conform with Alberta regulations. He said Aspinall himself never fenced the machinery that he was now demanding be fenced. Finlay said that based on Aspinall's work at Maple Leaf, "[H]e does not appear to be a man capable of exercising, or giving any opinion, or judgement, as to what is right or wrong about Boilers or Engines." He said in the letter that Aspinall told some of the Maple Leaf workmen that he "would go for me." The comment appears to allege that Aspinall would try to get Finlay into trouble or fired. Finlay claimed that the whole thing was petty spite over the firing of Aspinall, who had subjected him "to all sorts of petty annoyances, on the meanest, trifling and technical points." Finlay clearly expressed his frustration in the letter to Stocks and asked, in his conclusion, if he might come to Edmonton to show Stocks his boiler inspector's reports and correspondence, and clear up the matter once and for all.

Stocks's reply was dismissive. He wrote that he understood that Aspinall had been let go because the mine could not afford to pay his salary during the strike, and not for anything beyond that. Stocks continued with the veiled threat that various inspectors had made complaints since 1908 regarding ventilation in the mine, and that the company had promised on July 30, 1910 (before the Bellevue disaster in the nearby West Canadian Collieries mine) "that a fan would be installed and put into operation at the earliest possible date." Stocks said he understood that this had not yet been done, even in the face of complaints received from the miners' representatives about the ventilation in the mine. He continued that he was certain that Aspinall did not intend to subject Finlay to petty annoyances, and that his requests do not appear unreasonable or not in compliance with the Act. It would not be necessary for Finlay to come to Edmonton, Stocks concluded, because he had copies of all of the correspondence. Finlay thereafter remained silent and compliant in the face of Aspinall's requests. It seems likely that Finlay's anger arose from being told what to do by a man for whom he had little respect.

While the battle over the safety lamps at Hillcrest was apparently solved with the purchase of a second set of new lamps, and the resignation of miners to their use, the disagreement over the lack of a wash house continued. On November 23, 1908, the secretary of the union wrote to Stocks to complain that of the 200 men employed at the Hillcrest mine, "about 40 of them are using the dirty, filthy hole we have here to wash in. Others cannot use it." But Hill had dug in his heels, still embittered over what he believed was unfair treatment. The secretary also claimed that Hill had told them that despite their entreaties, the opinion of the health inspector, and the dictates of the law, he would not build a wash house.

On December 1, Stocks informed the union that Hill promised to have a wash house built to comply with the law, and that the inspector would visit the mine shortly to urge Hill to proceed on the construction.

Whatever happened in that conversation between Heathcote and Hill, it was more than Hill was willing to bear. He was frustrated at every turn by people who made demands of him even though it was he who had found the best coal in the Crowsnest Pass, he who had worked hard to find capital and to plan and develop the mine, and he who provided jobs and income. But it was never enough. They always wanted more, and often expressed their demands without courtesy or civility. Hill was obligated to answer his investors' questions about why the mine kept soaking up their money. He was clearly a man of independent spirit, completely at odds with the idea that he should ever be bossed around. He had no magic wand to make a wash house instantaneously appear, and anyone associated with construction knows the incredible difficulty of predicting job completion dates in boom times, particularly in the face of a shortage of capital and the absence of skilled workers. Hill's need for development capital had magnified his management difficulties, and the stress had begun to take a toll on his health.

Hill approached Walter Hull Aldridge, who headed up the CPR's mining and metallurgy department and was also managing director of the CPR smelter at Trail. Back in late May of 1907, as part of his effort to raise capital, Hill had written to Aldridge to offer him a block of 2,550 shares for the price of $191,250, along with a full explanation of how the money would be used. Apparently Aldridge had turned down the offer, but the situation had now changed — for Hill as well as the mine. In the absence of adequate capital, the stresses had become too much for him. The first man he told was Aldridge.

CHAPTER 4
HILL SELLS

On December 19, 1908, Hill wrote a letter to Aldridge that he must have found difficult to put to paper. In effect, it was a surrender in the face of financial difficulties and a recalcitrant union. Hill's decision would have far-reaching implications. He told Aldridge that he had decided to sell the mine. He also reminded Aldridge of his own promise that if he decided to sell, he would give Aldridge the first opportunity to buy it. Aldridge considered the matter seriously and quickly. Whatever management problems Hill faced, the Hillcrest coal was valuable, and Aldridge owed the CPR a reliable source of quality steam coal in the wake of his failure at Bankhead and the untenable problems at Hosmer. Thanks to the battle among Hill, the mine inspector, and the union, Aldridge had been given an opportunity to make good on his mistakes, so he got to work.

First, Aldridge needed an expert to evaluate the property, so he arranged for Lewis Stockett, the general manager of the CPR's Bankhead and Hosmer mines, to visit the Hillcrest mine. An

American born in Pennsylvania, Stockett had entered the office of a mining engineer at age 14, and had advanced rapidly through various positions at different coal companies in the United States.[1] In 1904 he became the general manager of the Great Northern Railroad's mines at Great Falls, Montana, but the next year he accepted a position with the CPR as general superintendent of the coal mines branch of the Department of Natural Resources, with headquarters in Calgary. It was a coup for the CPR managers to lure a man of such qualifications away from their longtime rival. Stockett wound up a member of the American Institute of Mining and Metallurgical Engineers, and eventually a member of the Canadian Institute of Mining and Metallurgy. His energy in Canada also led him to become a member of the Calgary Board of Trade. When Aldridge asked Stockett to evaluate the Hillcrest property, he had asked one of the best minds in the business.

Three seams of coal lie on the Hillcrest property, but Hill worked only the upper, no. 1 seam. It lies like an undulating blanket fourteen or fifteen feet thick that slants down to the west. From the outcrop at the surface at the side of Drumm Creek, Hill had dug his main entry where coal was brought out of the mine. The no. 1 and no. 2 slants followed the seam downward, before branching off into tunnels measuring ten by nine feet. These were also called levels, as they were more or less level, with only a slight downhill grade towards the slant to assist the horses as they pulled the heavy, coal-laden cars to the slant. The no. 2 slant near the surface was often referred to as the "mud tunnel" because of its wetness: in the early days, Drumm Creek would flood into the slant during the spring melt or during heavy rain. The rooms were driven directly up the pitch from the levels, twelve feet wide with cross-cuts from one room to the next every fifty feet. Miners at the face shoveled the coal into a steel-covered chute, down which the coal was supposed to slide to the mine car waiting below on the level. When the coal became stuck in the chutes, which it often did in the Hill Crest mine on account of the shallow angle, men called buckers would kick, pry, and shovel it loose.

The working system of the Hill Crest mine was known by the industry term "pillar and stall." Miners would eventually extract, skip, or rob the support pillars, with the result that the roof would eventually collapse as planned. That part of the mine would then be abandoned and closed off. The extraction of coal by miners at Hillcrest demanded safety-conscious men, tight management, co-operation, extensive knowledge of mining technique, and practical experience.

In "A Report on Hill Crest Coal Mine" for W.H. Aldridge, Stockett wrote that miners were paid fifty cents per ton, used locked Wolf safety lamps, mostly lived in shacks, and that "there does not appear to be a great deal of mine timber on the property." This last comment was a statement of the obvious. Hillcrest was often referred to as "Stump Town." Forest fires, as well as the cutting of timber for construction and mine timbers, had reduced the forest around Hillcrest to stumps and blackened timbers. Sparks and, later, superheated steam from locomotives often ignited forest fires, which were common in any forested area in the Canadian West near railroad tracks, and posed extreme danger to nearby towns built mostly of wood.

Stockett reported that the main entrance to the mine was through a mud tunnel that had proved difficult to maintain in wet weather. The mouth of the slope opened alongside Drumm Creek, which Stockett reported was likely to pour down the slope in high water and drown out the lower workings. At that time it was the only entry to the mine from which coal could be hauled; later, however, after the new owners had blasted out the rock tunnel and the no. 1 entry, it became known as the no. 2 slope, or no. 2 mine.

Stockett also reported that a ten-ton Porter compressed air locomotive and Rand high-pressure compressor were at the mine site, but not yet installed. Hill's plan had been to use the compressed air motor to replace spike teams along the level. The report added that the town had a good hotel owned by Hill, and a store owned by his father-in-law. Stockett valued the property at $308,000, but about $300,000 more, he estimated, would have to be spent on upgrades. More money would need to be spent to take up options on nearby coal lands. The $300,000 would pay for houses for employees, the installation of electric light works, water and sewage systems, mine development, and equipment purchases. It would be expensive, but the new owners, whoever they were, would wind up with more than five million tons of the best locomotive steam coal in the Crowsnest Pass squeezed into a seam twelve to fourteen feet thick. All they had to do was get it out of the ground and find someone to buy it. That some-one would be the CPR.

Hill's annual report for 1908 outlined some of his difficulties of the past year. During the first quarter the mine worked only fifty-seven days, and was idle for thirty-four. Through the second quarter, a money panic struck the West: people were afraid to invest, the coal markets fell off, and many of the mine's orders were cancelled. In May, the CPR had to deal with many severe washouts on the tracks in the Crowsnest Division, which dras-tically reduced the supply of cars and forced a ten-day shutdown. Further uncertainty about the prospects for crops also affected output and the market, leading to a reduction in production. The third quarter opened well, but was spoiled again by car shortage, which meant the mine worked only sixty-three days. The car situ-ation improved only in the fourth quarter, and production rose as a result. Hill declared that the Hill Crest Coal and Coke Company employed an average of 202 men each day, with the mine in oper-ation for a total of 172 days. The wages paid out totaled $149,477.58.

Hill, meanwhile, resumed his hunt for development capital, which the mine required immediately, even in the face of his plan to sell. He asked J.S.C. Fraser, the manager of the Bank of Mont-real at Trail, for a personal loan of $95,500 and a loan to the Hill

Crest Coal and Coke Company of $153,000. The collateral for the two loans was to be seven-eighths of the company's stock, and all of the company's bonds, respectively. Fraser had previously sent a memo to Aldridge asking his opinion as to the safety of such loans. It seems likely that Hill had also used Aldridge's name as a reference. Aldridge suggested to Fraser that the directors of the Consolidated Mining and Smelting Company of Canada, owned wholly by the CPR, would, if the CPR did not object, guarantee Hill's advance if Hill would agree to give the Consolidated Company a favourable option for three years for the purchase of the property, or the control of the stock.

The year 1909 brought changes for Hillcrest and the mining industry in general. Hill's nemesis, Heathcote, was temporarily replaced with a new district inspector of mines, Robert Livingstone. And, of far greater significance, an eight-hour workday became law in Alberta.

On February 9 and 10, Livingstone inspected the Hillcrest mine and reported that it produced 500 tons per day with seventy-five men on day shift and twenty-six on nights. Significantly, he also reported that during his visit, he found the mine clear of gas, but that the fire bosses reported in their books that gas had been found recently in several places. This was not unusual in such a mine. Dynamite was still used for shot firing.

Livingstone also reported that the extraction of a number of pillars had begun, but stopped because of a disagreement between management and miners regarding the prices to be paid for this class of work. This had become a huge source of frustration for Hill that would soon erupt into a full-blown labour dispute.

Furthermore, the issue of the wash house would not go away. Livingstone was critical of Hill for the lack of a new wash house, as Heathcote had been before him. The only place provided for miners to wash, he explained to Stocks, was directly over the boilers in the power house. The smokestack passed through the floor, leaving an open space around the stack of about ten inches. Dirt and soot from boiler tubes came up through the opening around the stack into the dressing room. The men employed in the mine, he said, could not conveniently wash, dry, or change their clothes. Livingstone pointed out in his letter that General Rule 26 of the Mines Act stated unequivocally that mines must provide sufficient accommodation, which shall not be in the engine house or boiler house. During his visit to the Hillcrest boiler house he had seen miners' pit clothes hanging along the steam pipe over the boiler. The mine manager, Cory Weatherby, had told him that the men hung them there to dry. Livingstone informed Stocks in the letter that he planned to write to Hill, and order him to provide a new wash house. He did just that, and pointed out to Hill the danger that miners faced with the current dressing room, should an accident occur to the steam header or the steam fittings.

Gossip, meanwhile, had begun to circulate in the Pass about the future of the Hill Crest Coal and Coke Company. An employee of West Canadian Collieries, J. Dougall, saw opportunity and wrote several letters to Aldridge about Hillcrest. He reported in February 1909 that he had "made a little trip to Hillcrest Mines yesterday to see for myself what was doing. Found the most disgraceful set of affairs." Dougall had ridden to the mine on a CPR locomotive to get some loads, but found that he could not get to the tipple because the Hillcrest locomotive was frozen on the tracks. The CPR engine had to return to Hillcrest station without a load. Dougall found the tipple shut down as well, and was told that the manager had resigned, but he was unable to find out who was in charge. Everything, he wrote to Aldridge, was disorganized.

Then, in April, mines in the Pass shut down again when District 18 miners walked out on strike over a new contract. The UMWA decided to fight for a closed shop because it would greatly increase union power and crush opposition among miners. The bitter strike pitted the president of the Western Coal Operators Association, Lewis Stockett, against the radical socialist president of the UMWA, Frank Sherman.

On the first day of the 1909 bargaining talks in MacLeod, the Crowsnest Pass Coal Company, Maple Leaf Coal at Bellevue, and West Canadian Collieries at Bellevue pulled out of the Western Coal Operators Association to negotiate separate deals with the union. Sherman got the deal he wanted from Crowsnest, which included a closed shop, but the agreement signed at MacLeod, negotiated in his absence by the union with the remains of the Western Coal Operators Association, did not include a closed shop. Despite that, the membership passed the agreement. Sherman, however, saw the lack of a closed shop concession, of the kind he had won at J.J. Hill's Michel mine, as a defeat for trade unionism, so he reneged on the MacLeod agreement. The April 1 issue of the *Lethbridge Herald,* a newspaper owned by W.A. Buchanan, a "Laurier Liberal" sympathetic to the UMWA, quoted Sherman as saying, "We have the Crows Nest behind us; The Maple Leaf Coal Co. and the Canada West will sign up and then we will force the Western Coal Operators on their knees and the Western Coal Operators are the CPR. We will force them into line." Despite his bold words, the strike dragged on for four months.

During the strike, Hill's personal battle with the UMWA came to a head. Hill wanted to pay miners less to extract coal from pillars than the union expected him to pay. Hill claimed, and the industry generally agreed, that it was easier to extract coal from the pillars than from the face. The Joint Committee, a dispute resolution mechanism consisting of representatives of the Western Coal Operators Association and the UMWA District 18, convened on Wednesday, June 2. The first order of business was a visit to the mine so that the Joint Committee could see the state of the operations firsthand, as well as the conditions under which the men worked. In the afternoon, the Committee called witnesses who were sworn in and questioned by Joint Committee members, both company and union. Although the Joint Committee was assembled to deal with a single specific complaint, the testimony at the hearing paints a revealing portrait of Hill the man, and the struggles and frustrations he faced as the owner of a small, undercapital-ized mine. Among the more eminent names present were Lewis Stockett, president of the Western Coal Operators Association, and Frank Sherman, president of District 18. Sherman, gravely ill with kidney disease, would be dead within five months.

Under questioning, Hill reminded the panel that he had many extra expenses that other mines in the district did not, such as a mud tunnel, a long haul by rail from the pit mouth to the tipple, a retarding conveyor (to get coal down the hill to the cars), and a standard gauge railway, locomotive, and crew to get the coal to the CPR tracks. Hill said it cost him $0.14 more per ton to extract the pillars than it did to mine from the face, and $0.20 more per ton to handle coal after he got it in the chute than it did other mines in the Pass. Hill asked for a reduction of $0.15 per ton on pillars, although he said the reduction should really be $0.20 or $0.25 per ton.

Hill complained that promises made by the union before he signed a contract — specifying that he would not have to pay more than $0.30 per ton for pillar coal — had not been kept. Both sides had agreed before the contract was signed, two years pre-viously, that the union would give him a price for pillars later. When the time came, however, the union refused Hill's request for a price, and threatened to shut him down if he went ahead without their approval and mined the pillars for less than $0.50 per ton. Hill had begun to pay $0.40 per ton without an agreement, but the men refused to work. As a result, he lost a contract when he could not supply the coal.

Hill also took issue with the price he paid men to install tim-ber props. He had agreed before the contract was signed to pay $0.05 per lineal foot for the installation of heavy timbers in the main entry, but now he was paying $0.05 per foot for even the very light timber in the rooms. Contrary to Hill's instructions, the contract had been signed by his superintendent in his absence, and the union local had interpreted it as meaning $0.05 per foot even for the small room timber.

From Hill's point of view, he faced an attitude problem with many of the miners. Hill claimed that bad mining practice had resulted in a great deal of dirt and rock in the coal, and it put the mine in danger of losing another important contract. He recounted a conversation with the men whom he had begged to mine the coal with more care, because it was the only contract he had left. Hill asked them if they would mine the coal "like men and give me a fair deal and I was answered that they would do as they damned pleased, and that they would stay at Hillcrest as long as they please, and that no matter what I said they would stay there as long as they were members of the union."

Hill explained to the Committee that he was certain he would have to reduce the selling price of his coal. For one thing, he said, there were too many mines in the Pass for the market. In response to a question from Sherman, Hill said that "The CPR are [sic] a great power, and with the Union to push you up on one side and a big railway to pull you down on the other, you had better get out of the middle. I have tried it for five years and you drove me to a shelter. The formation of a Union has crushed an individual concern from business."

In response to another question, Hill continued, "No mine can run in this pass in my opinion without a contract from a Railroad Company. The consumption of the domestic trade doesn't touch these mines. We are bound to the trade of steam."

Hill explained that the prairie market for coal was unstable and unreliable. "[M]y books will show you that over three percent of coal shipped to the prairie is a dead loss to the mine — that you can't collect the price paid the miner and the operation of the plant, you can't collect enough money from these people to pay for the actual cost of the coal. The losses are more than you can imagine. A man orders a car of coal in cold weather by wire, and even the Government agents will back him up — you ship the car of coal and you do everything to get your money — he tell you when he plow he will sent [sic] it but you don't get it — when he reaps his harvest he will send it, but you don't get it. You must deal with the railroad companies and you must take their prices, or you must lie idle."

Hill was asked again about rock in the coal. He responded that he was told by the miners "with malicious intent — we will put this rock in this coal and you can put picking tables and pick it out." He was asked if he was sure this was said maliciously. "I maintain that the degree of bad Socialist here has taught the men to think that they own that mine — they practically owned that mine and did as they pleased regardless of management." Hill said he had lost all of his contracts on the books during November and December of 1908 because of dirty coal. By 1911, the contract between the union and the Operators Association specified that after a first warning about rock or other impurities in the coal given to a miner, a second offence could result in discharge. By then, it was too late for Hill.

Summing up his frustration, Hill said in response to Sherman, "I feel this: that I have tried very hard for five years to get this mine working, between trying to get these men to work at this end — I have been many a time to all union officials and tried the best I knew how to get something established that I could depend upon whether the mine could be worked or not. I never felt there was three months ahead of me since I have been running this mine that I could depend upon at this end — a very hard position."

Sherman then asked Hill, "You seem at this time to put all your troubles on the miners and mine unions — I notice that you have very few men here that you had a year ago?

Hill replied, "I want it distinctly understood I don't blame the men of this camp — I blame a dozen agitators for this trouble."

"But you got rid of those?"

"I did some of them, but it was an awful task."

The Joint Committee ruled in Hill's favour, and in so doing, the committee's chairman pointed out, "[W]e must remember that money is put into Coal Mines for profits; and unless profits are the outcome of operations money will seek some other channel in which it may adequately get its increase." It was a statement of basic economics, whose truth was obvious even to the union members of the committee. They agreed with Hill's position on every major point, and cut the price on pillar work to $0.40 per ton and on props to $0.04 per lineal foot.

By September 30, another district inspector of mines, John T. Stirling, reported that "a number of men are employed putting in the concrete foundations [of a new wash house] and it is expected that this house will be ready for the use of the workmen within the next two months."

The bitter war of words between the union and the operators continued as the strike dragged on, but the conflict also led to internal union disputes between District 18 and the International Board of the UMWA, which insisted, as did Stockett and the mine operators, that the MacLeod agreement must be honoured. The stalemate between the owners and the miners finally ended with the imposition of a federal disputes resolution. With the strike dragging on and winter approaching, the government convened a board, and urged the union to drop the demand for a closed shop. The strike finally ended on July 19, 1909 without the closed shop concession that Sherman had demanded. For Sherman it was a bitter loss, and for the Western Coal Operators only a temporary victory. Eventually the union would get their demand.

While the mine operations continued at Hill Crest Collieries in the wake of the settlement of the Crowsnest strike, the due diligence inspection of the mine was undertaken by none other than Lewis Stockett, general manager of the coal mines for the CPR. In late August 1909, he again visited the mine, this time to inspect its accounts at Aldridge's request. The company's books, Stockett reported, had been kept in a very lax manner before the arrival of the current accountant, Cochrane, near the end of March. Until Cochrane arrived, there had never been monthly statements of costs or balances. Because the books were in such pitiable condition, Stockett said that he was unable to go into things in great detail. Stockett only had access to the accounts for one afternoon, in Hill's presence, after which Hill had left on other business. Stockett revealed that Cochrane, who had been an employee of the Crows Nest Pass Coal Company at Fernie before he was employed at Hillcrest, had, contrary to Hill's wishes, sent a statement of Hill Crest's costs of operations and average prices received for coal, along with other information, to Aldridge. In conclusion, Stockett suggested that even a complete audit might not reveal the true standing of the company.

The CPR had become deeply involved in the purchase of Hill Crest Collieries, not only through the involvement of Aldridge and Stockett but also with the explicit knowledge of their involvement by their president, Sir Thomas Shaughnessy, who "had no objection whatever" to Aldridge purchasing a small amount of stock.[2]

Aldridge had quietly assembled investors to buy the Hillcrest mine. On December 3, 1909, he wrote a letter offering the management position to John Brown, a Canadian with mine experience in the Crowsnest Pass, who was then employed as manager at the Bolen-Darnall Coal Company in Hartford, Arkansas. Brown held mine manager's certificates for Oklahoma, Illinois, Alberta, and British Columbia. Brown had been the superintendent at the CPR-owned Hosmer mine, and had worked his way up through the ranks. He had once been responsible for the Bankhead mine, and had driven a main tunnel there. He earned the respect of Aldridge, who considered him energetic and likely to spend considerable time underground — a trait that Aldridge considered an asset for good management. Aldridge thought Brown, although not particularly well-educated, would prove to be a disciplinarian, which he considered important in coal mining. The only other candidate considered, Robert G. Drinnan of Vancouver, was much better educated than Brown.

In the letter, Aldridge told Brown that friends of his had an option on control of Hillcrest, and offered Brown the management position for $5,000 per year if it were exercised. He asked Brown when would be the earliest that he could reach Hillcrest upon acceptance of the offer. In a later letter, just before Christmas, Aldridge said he had heard nothing further with regard to the Hillcrest property, that he had advised the Montreal investors not to proceed further unless they secured legal title to Hill's homestead, where his house and the entrance to the mine were located, and that Hill should not be permitted to retain any property or residence. "If he retains his residence," Aldridge wrote, "he will live there most of the time and will be an absolute nuisance to the new management." Aldridge advised Brown to say nothing to his present company about leaving. At that point in time, the sale was by no means certain.

Less than a week later, J.M. Mackie, who headed up the new ownership group, received an extensive report from Stockett based on his own examinations of the properties. It included a complete description of the lands, the coal seams, the mine interior, the methods of mining used, the buildings and plant, haulage, ventilation, the labour situation, and a plan of action for the future of the mine. Stockett was obviously excited about the prospects for Hill Crest Coal and Coke. He had revised his estimate of the amount of recoverable coal substantially upwards. His new conservative estimate of the amount of easily recoverable coal was 10,000,000 tons, but in all likelihood the amount would be more than twice that just from the fourteen-foot-thick no. 1 seam. Two other seams had not even been touched yet. The prospects looked good — but there were problems.

Handicaps faced by the mine included an almost continual shortage of railway cars from the CPR to ship the coal to buyers, as well as poor labour conditions. To make it a first-class mine would require a new system of opening up the mine, and a great deal of new machinery and buildings at a cost of $300,000 to $400,000. The mine owned one locomotive, affectionately known as "Old Maud" among the miners, used on the spur from the tipple to the

CPR main line; a second small locomotive, the dinky, which took coal from the entry at Drumm Creek to the tipple 3,300 feet away; and an air compressor, along with a ten-ton Porter Air Locomotive which the company had not yet installed. Workmen had not completed the wash house, and many of the mine buildings were only temporary. As for the labour situation, Stockett described the mineworkers as strongly union. They needed to be handled with diplomacy, he said, if trouble were to be avoided. In any case, he urged a considerable margin to be allowed in cost estimates to mitigate strikes and other labour troubles. Accommodations for men were poor, with the result that Hillcrest did not get what Stockett described as "the best class of men."

Stockett reported a fair number of handicaps, but he also suggested improvements. He recommended the construction of houses that could be rented or sold on an installment basis. The result would be a good return on investment of perhaps $50,000 directly, and increased efficiency on the part of the mineworkers. Stockett reiterated the claim that Hill Crest coal was about the best of the Alberta coals and therefore highly marketable. It was being sold to the Canadian Pacific, Canadian Northern, and Milwaukee railways, as well as for domestic use at upwards of two dollars per ton. Among his extensive development suggestions, Stockett urged the construction of a new 1,200-foot rock tunnel that would become the main entry — exactly what Hill himself had planned to do. Stockett's numbers on projected future profits and the extensive report seemed to be what the potential buyers were looking for, but the new group wanted to break cleanly with Hill's management and present a new face to labour. Hill, however, was still involved in running the mine. Stockett's report to the ownership group convinced them to proceed with the purchase. Aldridge passed the news to Brown, who was still in Arkansas, in mid-December. Brown then sat down with his employer, Mr. Bolen of the Bolen-Darnall Coal Company, to give his notice. Brown described in a letter the deep effect of his plans upon Bolen. The grammatical and spelling errors illustrate his lack of formal education, and the way in which Brown's value to Aldridge extended beyond his education:

"The Old Gentleman, sat a long time without saying anything then he looked across the desk at me and all he said was John are You sure You are not making a mistake. Then he went out of the office and I did not see him until the next day, in the mean time he had gone down to the bank and talked the matter over with My Brother, he ask him to try and get me to give up the notion of going back into the North-West."

In the same letter, Brown revealed his dislike for Hill: "I think it would be best for all parties concerned even Hill himself. if he was removed from of the Board, and out of the place altogether with his Wife, and her relatives, for I used to hear that his Wifes relatives raised more H- - l at Hillcrest, than Hill himself ever did at his worse." [3]

On December 2, 1909, Hill informed Deputy Minister John Stocks that James Sommerville Quigley had been appointed manager of the mine. Hill had made a good choice. Born in Fauldhouse, Scotland in 1875, Quigley had worked in the mines of Nova Scotia, and, like many young men, had been lured to the Klondike along with his brother Tom during the gold rush of '98. They had each packed their supplies up the golden staircase and headed to the Yukon gold fields, but the brutality of life as a prospector took a toll. Tom left the Yukon first, convinced that the hardships were too great and the chance of striking it rich too slim. James stuck it out for a time, and found some gold, but finally he realized, like Hill before him, that his future lay in coal. Quigley left with a few nuggets, some of which he strung on a necklace that his family cherishes to this day, and moved to Cochrane, Alberta, where he worked as a mine manager before Hill hired him at Hillcrest. The appointment of a man they did not know caused some anxiety among the group maneuvering to buy the mine, but when Quigley turned out to be a competent and fair man, they too were satisfied with the hire.

Despite Hill's desire to sell Hill Crest Collieries, largely out of frustration, he wrote a report entitled "How the Revenue of This Company Could be Increased." Among his suggestions to the ownership group was the installation of 200 coke ovens. This explained why Hill had named his company the Hill Crest Coal and Coke Company, and that his drive to play a role, to imagine a future, and to succeed had not diminished even in the face of his management difficulties. Hill had intended to build coke ovens from the very beginning, but his lack of investment capital had prevented it. To the new owners, the suggestions in Hill's report only showed that he intended to meddle. On January 5, 1910, the agreement was signed and the new owners took over, but Hill remained as one of the company's directors, as well as the second major shareholder behind the ownership group, C. Meredith & Company Ltd.

Dougall, the unsolicited spy, had given Aldridge faulty information, something Aldridge knew as soon as he read the letters. Hill's former general manager, Fraser, had either quit or been dismissed, and Hill had hired MacKenzie to replace him. James Cochrane, the new accountant who had started at the end of March and quickly became Hill's right-hand man, could not praise MacKenzie enough, and claimed that he was the general manager that Hill had needed for years. Cochrane reported to Hill on January 9, 1910 that MacKenzie was a doer, not a talker, and had made it clear to other staff that loitering and smoking were cause for dismissal. Not only that, he had sent the engineering staff packing and replaced them with competent men; he had also hired a new locomotive engineer and mechanic. Cochrane also reported to Hill that the fitting of the lockers in the new wash house had been a big job, but they were swiftly erected, thanks to MacKenzie. MacKenzie, however, would not remain long with Hill Crest. He quit, some months after Dougall's letters claimed that he had. MacKenzie's brief tenure had seen one important task completed: the wash house. By December 29, 1910, workmen had finished work on the new wash house, but it was not yet in use. The union complained again, and by late January, inspector Elijah Heathcote, back in the fray, reported that the wash house had finally been put into full operation. It had taken over a year to build and equip.

The effort to get rid of Hill as an influence on the mine management increased once the new ownership group took over. Mackie, the head of the group, told Aldridge in a letter on January 8, 1910 that Hill would give up the homestead property completely, and the new company would obtain possession of it from the government by a special dispensation. In other words, the ownership group had approached Ottawa politicians and managed to overcome the particular homestead regulations standing in the way of their complete control of the property. Hill's house and ground north of Drumm Creek would transfer to Hill, but subject to the new company's rights. As for Hill himself, he would shortly depart for a tour of Europe. In January, Aldridge wrote to Mackie that "[i]t will certainly be better for Hill and for the company as a whole if he leaves the property for six or eight months. This will permit the new management to straighten matters out." Aldridge recommended that the new manager should not be installed until the transfer of the property that adjoined the Hillcrest company's lands was complete.

The manager-in-waiting, John Brown, who held off on his arrival at Hillcrest until Hill departed and the property transfer was complete, had his own problems to deal with: the mine he general managed in Arkansas had blown up. During his attempt to rescue a shot firer, Brown had been overcome with afterdamp and lay unconscious in the mine for 30 hours before being was found by rescuers. As Brown was rushed to the hospital, hot bricks were stacked around him — part of standard treatment at the time for victims of CO poisoning. But the jostling of the ambulance upset the bricks, and Brown was badly burnt en route to the hospital. The severity of Brown's injuries, and his inability to communicate with Aldridge, forced the ownership group to briefly consider their second choice for the manager position: Robert G. Drinnan of Vancouver. Significantly better educated than Brown, Drinnan was also, in Aldridge's opinion, more of an engineer and office man, but would produce coal at greater cost than Brown would because of Brown's greater willingness to supervise the workmen underground. Brown's personal troubles, however, were not over yet.

In mid-February 1910, Brown and his wife lost a child. Brown buried the baby himself. Once his concerns about his wife's welfare were allayed, Brown finally wrote to Aldridge that she seemed to be "getting along fine and that is the main thing with me. I am feeling in very good shape again trusting things are going well with you." Word from Brown that he had regained his health was all that Aldridge needed to hear. Brown was the man.

On March 3, Mackie wrote to Brown again, this time to tell him that "[t]he Hillcrest Company is not controlled by the Canadian Pacific, the Consolidated or any of the Companies with which I am associated." This odd declaration was likely intended to defuse rumours that refused to die — even as late as 1914 — suggesting that the CPR was the actual ownership group. Mackie added, on another matter, "I understand there has been some slight friction between the new controlling interests and your friend C.P. Hill. I want you, however to use as much Scotch [sic] diplomacy as you possess in keeping Hill friendly and avoiding trouble, without of course permitting him to in any way interfere with or injure the company's interests." Aldridge further warned Brown about the Hillcrest labour situation, and to take care in handling it when he got there. "I imagine there is a fairly bad crowd and that they will make trouble at the least provocation, especially if any of them feel that the new management is not particularly friendly towards them."

Also on March 3, Aldridge replied to the letters he received from Dougall at West Canadian Collieries in Blairmore. These letters contained information that Aldridge knew to be false even as he received them. Among the half-truths that Dougall reported was the claim that Hill had sold the mine to M.P. Davis[4] of Toronto. In fact, Davis was only one of the shareholders in the new purchase group, as well as one of Hill's original shareholders. Dougall commented in one letter that, "It looks as though C.P. [Hill] had handed someone a lemon, as there is no question but that the cream is off the property." According to Dougall, Hillcrest was having serious trouble with faulting; the quality of coal had deteriorated, and now was very "slacky" with considerable rock in it. Aldridge was well aware of the true value of the property, and of its importance to a great many people, including the CPR management.

Through his letters, and the passing on of supposed information, Dougall had hoped to grease the wheel and be offered the position of general manager. When Dougall finally revealed his motives and asked for the job, Aldridge explained that the position had been settled some time ago, and pointed out that many false reports about the ownership of the property had circulated. He also pointed out to Dougall, as he had to others, that neither the CPR, the Consolidated Company at Trail, nor "any of their allied Companies have any money invested in that property, or anything to do with its Management. Personally I occupy no position in connection with them, nor am I a director, although I understand such a report has been freely circulated." Aldridge did not mention that he held a small block of shares, although he would soon sell them off.

Shares in the new company were offered for sale in the *Toronto Star* on March 8, 1910. The large ad on behalf of C. Meredith & Company offered for public subscription 6,250 shares of seven percent preference stock, with a forty percent bonus of common stock in Hillcrest Collieries Limited. The directors included Hill and Brown of Hillcrest, along with H.S. Holt, president of the Royal Bank of Canada; J.S.C. Fraser, manager of the Bank of Montreal at Rossland; M.P. Davis, an engineer with Dominion Bridge Company of Montreal; W.D. Matthews, president of Consolidated Mining and Smelting Co. of Toronto; C.R. Hamilton, K.C. (King's Counsel) of Rossland; and J.M. Mackie, manager of Gould's Pump Co. of Montreal. The current presidents were C.B. Gordon, president of Dominion Textile Co., and Charles Meredith, president of C. Meredith & Co. Ltd.

The two big names on the board of directors were Holt and Gordon, influential Montreal businessmen who knew each other well. Both were directors of the Ritz-Carlton Hotel, and both were instrumental in the establishment of the town of Hampstead on Montreal Island. They developed it as a garden city with large house lots for the inclusion of trees and shrubbery, curved streets, and an absence of retail shops. Then, as now, it was the domain of the affluent.

Holt was formally trained as an engineer, owned or directed over 300 companies, and supervised the construction of the CPR through the Rocky Mountains. He owned the Montreal Light, Heat and Power Company, which supplied power to the entire Montreal area, and would allegedly control Montreal's tramway network as early as 1914. Fourteen years later, Holt's personal fortune was estimated at $3 billion, at a time when the federal government had only $300 million in circulation. But he had his detractors. When two CPR employees tried to murder Holt, he knocked them both unconscious. Another time he was shot by a shareholder, but survived.

Gordon, who was an acquaintance and business associate of Holt's, was also a man of vast wealth and power. Active at the core of the industrial elite of Montreal, he held an impressive portfolio of directorships and holdings: president of Montreal Cottons Limited, director of the Bank of Montreal and Canadian Car and Foundry, CEO of Dominion Glass and Dominion Textiles, Ogilvie Flour Mills and Royal Trust. Hillcrest Collieries, although not owned by the CPR, was thus controlled by men of vast wealth in Montreal who had intimate connections to the railway.

Hill, who still wanted to manage the mine, particularly now that it had become properly capitalized, travelled to Winnipeg in search of coal buyers, and then wired Mackie, the head of the new management group. Hill believed he could get a contract from the CPR to supply the railway with 500 tons of steam coal daily, at $2.10 per ton. Mackie in turn wrote to Fraser in Trail to inform him of the good news.

The intrigues were not over yet. On March 15, Aldridge wrote to Brown in Arkansas to warn him about the accountant, Cochrane. Aldridge claimed he was an impossible man for the position. Cochrane had been jockeying to keep his job, and had sensed the new management group's hostility towards Hill. He had attempted to make it appear as though he and Hill were in conflict with one another, but Aldridge was certain they were in fact very close. He warned Brown not to say anything on the subject when he got there. Aldridge wrote that he had discovered Cochrane was responsible for the resignation of the last manager, MacKenzie, and there was little doubt that he would try to make it disagreeable for whoever replaced him.

Aldridge then wrote to Mackie in Montreal to report on his communications with Brown, and on the state of the Hillcrest mine. Aldridge wrote that a fault in the coal seam was causing problems. He also gave a backhanded compliment to Hill: "I understand they have had some difficulty in keeping men at work. Hill deserves great credit for 'hanging on by the skin of his teeth' as long as he did, but the results of miserable management are now very much in evidence." On March 28, Brown wrote to Aldridge to explain that he had to deal with another explosion in his mine at Hartford, which had killed a shot firer. There had been several accidents in the Arkansas mining district, and in each case a shot firer had been killed. Brown had gone straight from the train station into the Bolen-Darnall mine to get out the body. This took nine hours.

Meanwhile, back in Hillcrest, Hill vacated his residence near the mine for good on April 6, 1910 and left to tour Europe with his family. Although still a major shareholder, he retained no role whatsoever in day-to-day mine management aside from his position as one of the company's directors. Hill's departure was the signal that the new owners had been waiting for. On April 8, Brown arrived at Hillcrest and strode into the management office of Hillcrest Collieries. The new man had arrived.

CHAPTER 5
THE BELLEVUE DISASTER

Brown's arrival heralded a new era in management at Hillcrest Collieries, but he quickly discovered that the situation was not as bad as he had been led to believe. The miners he met there, he said, were the same as anywhere. He had also decided, apparently, that the continued employment of Cochrane would not be a problem, because in mid-June Cochrane was still Hillcrest's accountant. Over the following four years under Brown's management, the bitter struggle with the union would largely disappear, thanks to his thrust to improve labour-management relations. The rancorous disputes with provincial mining inspectors would also dissolve into a hardly noticed routine.

One of Brown's first and best decisions was to employ William Hutchison as mine engineer and surveyor. Hutchison, born in 1874, brought an illustrious background to the mix. His father was a cousin of Sir Sanford Fleming, the engineer and surveyor who had been in charge of the initial survey for the Canadian Pacific Railway, and who first conceived the idea of time zones.

Educated as an engineer, Hutchison had worked, just prior to his hire at Hillcrest, at the cement plant at Exshaw, where he had designed the kilns for baking limestone, as well as a dam and powerhouse. Hill Crest Coal and Coke had supplied coal to the Exshaw cement plant for use in its kilns.[1] Now Hutchison became the brains behind major changes at Hillcrest Collieries.

One of these changes, pursued at Aldridge's request, would have far-reaching implications. Aldridge revealed to Brown on April 14, 1910 what he planned. He had urged the Hillcrest Collieries directors to secure from West Canadian the acreage they owned on Byron Creek. Stockett and several other experts had examined the property, and said they believed it was one of the very best coal properties along the Crowsnest Pass. The Byron Creek property adjoined the Hillcrest property on the south, and, according to Aldridge, the acquisition of these seams would give Hillcrest the equivalent of two new mines. The prescience of this initiative would not become clear until years later, when these seams would ensure the survival of the company and prolong its life. For now, though, it was important to get the coal out of the Hillcrest mine.

The new company wanted to hike production to at least 2,000 tons per day, which would involve an expansion of the underground workings. As well, the changes would require power for haulage, ventilation, and lights. As soon as Hutchison arrived, on May 10, 1910, he began work to sink two new slopes, or slants, and the new rock tunnel recommended by Stockett, which would give the coal a second route to the surface and eliminate the need for expensive air locomotives for haulage.

Arthur Watson, an Englishman from Derbyshire and veteran of the Boer War, arrived in Canada at Frank in 1905 and worked with the crew that drove the rock tunnel at Hillcrest in 1910. Watson and the other men who drove the tunnel used dynamite for the blasting, and that did not sit well with mine inspector Heathcote. On May 26, 1910 he wrote to Brown and insisted that the use of dynamite was not in accordance with the Mines Act, instead recommending the use of monobel saxonite. Brown replied that he had read and reread the Mines Act, and could find nothing to indicate that the use of dynamite was prohibited, but promised that he would try monobel as soon as his stock of dynamite was used up.

In November, water flooded the mine, and the lack of electric power to drive the water pumps forced Hutchison to stop work on the tunnel. Rather than cease the development work, however, Brown focused on driving through the new slant, which would connect with the rock tunnel.

It was at this time that Hill briefly returned to Hillcrest. Brown, who intensely disliked him, wrote, tongue firmly planted in cheek, "My friend C.P. Hill, arrived in Hillcrest on last sunday morning. [I] listened to a lot of good humoured hot air from him on sunday and monday since then I have not seen anything of him." Just as the rock tunnel neared completion in December 1910, events at a nearby mine attracted the attention of Crowsnest miners. The West Canadian Collieries mine in Bellevue blew up.

The story of the Bellevue disaster is interesting for several reasons, not the least of which being the death of a rescuer named Fred Alderson.

Alderson was a native of Sunderland, County Durham, in England. His father drowned at sea when Alderson was young, and when his mother remarried, she placed him and two brothers in an orphanage. When he was old enough, Alderson left the orphanage and travelled to South Africa to drill wells. Later he returned to England and took up coal mining, studying mining and engineering at night before leaving to work in India for a year. A later venture at a coal mine in Mexico failed, and he decided to come to Canada, where he got a job with his brother at Hosmer.

Alderson became a popular fire boss at Hosmer and volunteered to join their rescue squad after the B.C. Department of Mines ruled that all large collieries must equip themselves with mine rescue apparatus and trained rescue teams. Shortly afterwards he applied for a position as district mines inspector in Alberta. Alderson received his rescue training and was called to Alberta, but not as a district mining inspector. Within weeks of the formation of the first Draeger Mine Rescue Team in the east Kootenays, the mine at Bellevue blew up at seven in the evening on December 9. Alderson was called as a member of the rescue team.

Typically, following a coal mine explosion, the miners wander the workings in search of a way out. When the explosion occurred in the Bellevue mine, Andrew Matson was operating a compressed air drill underground.[2] Unable to operate the drill because of the sudden drop of air pressure in the feed line, he observed smoke carried in the ventilation air current for several minutes. Then, unaware of what had happened or of the seriousness of the situation, he sat down to eat his supper. After about ten minutes, Matson noticed that the ventilation had stopped. Now aware that his life was at risk, he and two other men tried to leave the mine by the usual route, but found their way blocked by afterdamp. Made ill and weak by the poisonous gas, they, along with three other men, retreated to an air motor and used the motor's compressed air to revive themselves from the effects of the carbon monoxide. Eventually, they retreated to the charge station, and stayed there all night with no ill effects. After eleven PM, the high-pressure air line resumed pressure. Eight men waited there for rescue, unable to escape because of heavy afterdamp.

Not counting the blast itself, the greatest killer in coal mine explosions is afterdamp, making speed essential in any rescue effort. At Bellevue, the same wall of afterdamp that prevented men from escape also prevented rescue teams from entering the mine to reach them. Bellevue's manager, John Powell, was well aware of Hosmer's new, fully equipped rescue team, and called

John Stockett, the general manager of the Hosmer mine, who moved quickly. He and the team — along with James Ashworth, the general manager of the Crows Nest Pass Coal Company's Fernie and Coal Creek mines, the Coal Creek manager; John Shanks, the Michel manager who had set up the first Draeger-equipped rescue team; Norman Fraser; two B.C. mines inspectors; and a few others — left immediately on a CPR special train that rushed toward Bellevue through the bitterly cold December night.

In the minutes after the explosion, mine manager John Powell, Frank Lewis, and pit boss John Anderson crouched low to avoid the afterdamp and ran into the gassy zone in a daring rescue attempt. They found two men alive and two dead, but the afterdamp was so thick that it put out Powell's lamp and made them all sick. Still, they managed to drag out the two surviving men. One had been struck in the head by a piece of coal moments before the explosion. His wound bled copiously as he tried to escape, but overcome by afterdamp and loss of blood, he lost consciousness. Powell and Lewis dragged him to a place where the air was good, and ran back for the next man. Once Powell had recovered from his own CO sickness, he and other rescuers began work to restore the brattice, stoppings, and ventilation disrupted by the explosion, while the fan, which had never stopped, continued to operate. Their determined but slow restoration of the ventilation allowed them to advance gradually into the afterdamp zone.

Soon the rescuers saw the lights from safety lamps farther down the tunnel. Once again, Powell ran with three others into the afterdamp. This time they found twenty-one men grouped around a charging station with the pipe broken off — the source of the pressure drop in the high-pressure line. In groups of four, without any breathing apparatus, rescuers dashed into the thick afterdamp at great personal risk, and pulled out all twenty-one men. All of these men were dead, but others in the mine might still be alive. The brattice teams had restored ventilation close to room 86 when B.C. Mines Inspector Strachan arrived from

British Columbia with the Draeger apparatus at about 2 AM. The Hosmer team brought four sets of two-hour apparatus and several half-hour units with them. The Draeger unit worn by each member of a rescue team was a self-contained apparatus with an entirely lung-governed oxygen feed that, while heavy and ungainly, allowed rescuers to safely enter the mine's poisonous carbon monoxide atmosphere.

Strachan and the other trained rescuer, Alderson, wearing the Draeger breathing apparatus, advanced through the zone of afterdamp to find a group of men alive and grouped around the air recharging station for air locomotives. This was the group of men that included Matson and fellow miner Frederick Heal. Alderson took off his own apparatus, handed it to Heal, and helped him to put it on. Alderson stayed at the charging station while Strachan left. Once Strachan had led Heal out of the zone of afterdamp and made sure he was safe, Strachan returned to the group carrying the breathing apparatus that Heal had worn. Then it was Alderson's turn. He put on the apparatus, left Strachan behind without apparatus, and rescued another miner the same way that Strachan had, then returned the group awaiting rescue. The rescuers expected Strachan, then, to bring out another miner, but he returned to the rescuers alone from the group of stranded men. He said that Alderson had dropped the 45-pound apparatus on his way back to the group.

Draegerman Matusky of Michel entered the zone of afterdamp next carrying a half-hour unit with him to bring out Alderson. When he reached the group, however, he found Alderson collapsed but able to speak, and the rest unconscious. Matusky left the spare apparatus behind, and hurried back to the other rescuers. He told rescuers that no one was able to put on the apparatus because they were weak or unconscious from the afterdamp, and insisted the men would perish unless they were rescued quickly. The rescuers, however, were now short of oxygen and would have to wait for supplies from the surface before they could return to the group of seven miners, including Alderson, who awaited rescue. Something had to be done immediately.

Thomas Spruston, from the mine at Michel, suggested that the rescuers advance ten feet apart into the danger zone with a rope that could then be used to pull out the unconscious men one at a time. The rescuers made a quick decision, and acted on it: they rushed in groups of four through the thick afterdamp without breathing apparatuses, and began to pull out the unconscious men. It didn't go well. When they got to the fourth man, Spruston heard a rescuer behind him shout out that he was "all in"— i.e., that he had been overcome by the gas and could not continue. More rescuers began to call out that they were done for. Four men, including Alderson, remained to be rescued at the charge station.

With all of the rescuers down, Spruston untangled the heap of men at the charge station, and arranged them so that the escaping air would blow over them. He felt no ill effects from the gas at this point, so he turned back to help his fellow rescuers, some of whom he found unconscious. Others, including the Bellevue doctor tried to walk, but could not stand. Spruston saw the physician fall and rushed to his aid, but as he tried to pull the doctor from danger, he felt his own legs weaken and he also went down. Spruston heard one man of the rescue party call out for help, and told him to try to make his way to the air leak from the high pressure line ahead. He tried to take his own advice, crawling over bodies to the air, then lost consciousness. Others, including the B.C. mine inspector Evan Evans, N. Huby, Brownrigg, and McAuley, managed to pull to safety the rescuers and the remaining men at the air station. These acts of bravery saved the lives of five men. Of the small group gathered around the air recharging station, only Alderson and one other Bellevue miner could not be revived.

Rescuers expected that the men who were left behind, including Alderson, would be safe, with the compressed air helping them breathe. What happened remains a mystery. The air compressor continued to operate, contrary to some later accounts. The final death toll in the Bellevue disaster was thirty miners and one rescuer, leaving behind twenty-one widows and forty-two orphans. Alderson, a Mason, was buried at Hosmer on December 13, after a funeral service attended by hundreds.

Neither the coroner's inquest held in the Bellevue Union Hall nor its conclusion was straightforward, and each could have served as a warning to Crowsnest Pass operators and miners. The transcript of the inquest runs to 850 pages, a massive document that details an exhaustive inquest by coroner Pinkney.

The inquest began with a false start. The jury and Pinkney viewed the bodies, then began deliberations. Two jurors, however, suddenly stated that they objected to a full inquiry into the cause of the explosion. The foreman of the jury, Albert A. Cameron, who was also an employee of Hillcrest Collieries, refused to consider anything except the immediate cause of death of the victims of the explosion, and would not consider the source of the carbon monoxide that killed the men. When the representatives of the union and the provincial government then objected to Cameron's presence on the jury, Pinkney was forced to adjourn the inquest. He consulted with the provincial government, and on their advice appointed a new jury and began to hear testimony on December 19, 1910.

The inquest brought a number of salient facts to light, among them that there had been an earlier explosion in the mine, that in late November gas had been reported in the mine, and that the Bellevue union local had met on December 1 specifically to discuss the gas, had complained about it to management, and asked for an inspection by the provincial district mines inspector. No inspection had yet happened when the mine blew up.

Of the last bodies recovered from the December 9 disaster, two of the three victims found in chute 52 had received severe injuries. One man was found with a hole in his skull and the brain gone. He died instantly from the injury. Another man suffered four leg fractures, fractured thighs and pelvis, and abrasions, but had in fact died from carbon monoxide poisoning rather than his severe injuries. Two of these men also suffered from burns to their faces and hands. Miner George O'Brien testified that he actually considered that the flesh had been cooked on their bones. Most importantly, several men who had helped recover, strip, and wash bodies in the wash house afterwards testified that they found smoking materials on these same men. O'Brien said that they found between a dozen and twenty matches in one man's pockets. Fire boss Charles Chestnut testified that he had taken a watch, a key, and three or four matches from one man's pockets. Another man had a jackknife and matches. James Allsopp also testified that he found a pipe, matches, and tobacco on one of the three victims, and matches and a pocket knife on the man whose hands were burned the worst. He gave the pipe, tobacco, and matches to his manager the next day, who in turn presented them to the inquest.

The last man found by recovery crews was buried under as much as two feet of soft coal. Powell testified that he found the man in a sitting position, with the lamp between his legs, similar to how one might sit if cold, to gain warmth from the lamp. This was one of the men on whom matches had been found.

The district mine inspector was asked if sparks and an explosion might be caused by rocks striking each other with methane in the air. Heathcote dismissed the possibility.

The juryman pressed him. "You have never heard of that at all?"

"No," replied Heathcote. He also said that it would take sustained, heavy sparking to ignite gas. The question was asked to determine if the rock fall that occurred in the mine had caused the explosion, or was a result of it. It was rare for sparks created by rocks striking to cause explosions. Friction sparks were attributed as the cause of only 0.54 percent of mine explosions in the United States between 1851 and 1956.

Smoking was widely believed to have ignited the methane in the Bellevue explosion, a theory completely at odds with the official cause. The Bellevue inquest concluded that thirty men died from carbon monoxide poisoning, and one had been killed when his head was smashed by falling rock. It concluded furthermore that the carbon monoxide was caused by rock caving that compressed air below the roof as it fell. The compressed air allegedly rose in temperature to the point that it incompletely combusted the coal dust. But this conclusion was not unanimous. One member of the coroner's jury signed a rider to the effect that he agreed the men had died from carbon monoxide, but not that the carbon monoxide was created in the mine by percussion. The percussion theory was explained at the inquest by the general manager of the Crows Nest Pass Coal Company's Fernie and Coal Creek mines, James Ashworth, who argued that it was believed to have occurred before, and he was "certain" that it had occurred at Bellevue, too. The theory postulated that if rock fell from the roof over a large area of the workings where the pillars had been drawn, then the sudden fall of rock in the enclosed space would compress the air, which would in turn generate heat sufficient to coke the coal dust and generate carbon monoxide. As coal coked at a lower temperature than methane ignited, he believed the Bellevue disaster was not a gas explosion at all. Ashworth's theories absolved anyone of any responsibility for conditions leading to the disaster.

Ashworth was knowledgeable and persuasive, but wrong about the nature of the disaster, despite his effective sales job to the jury. As the chief Alberta mines inspector John T. Stirling would write later, "[T]here was no evidence to support such a theory, even if such a mode or origin were possible, and as some of the instances quoted in support of the theory were so inaccurate, and the references so misleading, it would be a waste of time to quote even the reasons advanced in support of such a theory."[3]

Ashworth carried his percussion theory even further and

speculated that a second fall of rock pushed firedamp down to the main counter and suffocated the men there, including Alderson.

A much more likely explanation, however, is that the air became contaminated by the afterdamp caused by the movement of air and gas when miners restored the stoppings and brattice. This movement would have pushed the afterdamp into the miners' faces. This theory is given credibility by Bellevue superintendent John Powell, who reported that Alderson told him, moments before he entered the zone of afterdamp, "[D]on't put up any more stoppings it is backing the damp up on them."[4] Compressed air, while it could have saved several men, would likely have proven inaccessible for seven who might not have even noticed as the air became more contaminated with the odourless afterdamp.

Even now, little is known about percussion as a cause of mine disasters. In 1960, at the Coalbrook Colliery in South Africa, a massive rock fall underground caused a percussive blast, and a similar blast happened in 1961 at the Wa Jing Colliery in China where 163,000 square metres of roof collapsed. At Bellevue, the roof collapse was nowhere near the area of rock falls that are known to have caused percussive blasts. In the absence of more hard facts about percussion, Ashworth's conclusion as to the cause of the disaster can be considered no better than speculation. He insisted in his testimony that the men had not been smoking, and that the discovery on them of one match or a thousand made absolutely no difference. But a methane explosion only takes one match, and any evidence of actual smoking would be destroyed or scattered in the blast. Still, he was right: there was no solid evidence that smoking or a lit match had led to the blast.

The wildly unlikely official explanation of a percussive blast, however, had strained to the limit the credulity of chief inspector Stirling. Many men would simply have accepted the official explanation, and let it go, but Stirling's character would not let him rest. He knew that a percussive blast had not killed thirty-one men, and he set out to find out what had.

Stirling released the report of his detailed investigation on November 7, 1912, nearly two years after the Bellevue disaster. He and Professor John Cadman based their report upon their probes of three different explosions in the mine: the blast on Thanksgiving Day 1910, the explosion on December 9, which killed the thirty-one men, and a third on approximately January 28, 1911. (The exact time of the third explosion is unknown because a general strike by miners meant no one was inside the mine when it occurred.)

A careful examination of each explosion revealed that the location of a fall of rock was also the geographic origin of the blast. Furthermore, the rock falls occurred in locations where firedamp was likely to accumulate due to poor ventilation. The somewhat disorganized method of mining in Bellevue made ventilation in some places very difficult, if not impossible. Stirling found ample evidence of flame on the props that completely discounted the percussive theory. The question then remained whether a fall of rock in the Bellevue mine could spark sufficiently to ignite fire-damp. The authors tried a simple experiment in which they lifted and dropped a large rock of sixty or seventy pounds into one of the chutes where some of the fallen roof lay. They observed a brilliant display of sparks. They also found that when a specimen was struck with a hammer, the surface of percussion glowed red hot for a moment. In further experiments Cadman struck samples of sandstone together, producing sparks intense enough to ignite coal gas and methane. The momentary combustion of the bitumen embedded in the sandstone provided the temperature required to ignite the methane. Despite the certainty that falls of rock very rarely caused explosions of methane in mines, Stirling and Cadman demonstrated that in this particular mine, that was exactly what had happened.

One interesting revelation of the inquest with wide-reaching implications surfaced from the interview between the lawyer for the province, William Campbell, and the secretary of the Bellevue union local, James Burke. Campbell asked Burke about the education of miners about the rules specific to each mine, and the regulations and laws that governed coal mining in Alberta:

"Do you know if any translation of these rules are made and put up in any way?"

"Not that I know of."

"Do you know what means there is of communicating from the English speaking operators of the mine and their hired assistants with the foreign speaking miners who come and cannot talk to them?"

"None whatever, as far as I am aware."

"There must be some way of giving them orders. They cannot find them by signs. Do many of them talk English or broken English?"

"Very few of them — I can't tell you exactly how many."

"But at your own meetings you do not talk foreign languages?"

"No, but we have interpreters."

"Are any of the officials for the mine interpreters?"

"No."

"Don't you know in connection with the mine whether these foreigners are told about the laws which bind them?"

"I am sure they are not told."

At that point, S.B. Woods of West Canadian Collieries took over the questioning.

"Is it the duty of the union, or of the Union officials — yourself, for instance, as paid secretary of that union — is it your duty to inform the members of the Union — these foreigners — that there are special rules — and to see that they get a copy?"

"No."

"You do not regard that as one of your duties to these people?"

"They are told it through one of their own countrymen."

"And don't you consider it one of your duties to look after their interests — to see that they get a copy of the rules and understand them?"

"We do our best."

"How do you do your best — by getting copies of the special rules?"

"No."

"Have you ever gone to the management and asked for copies of the special rules and been refused?"

"No I have not. There are various things I have done. I cannot remember all those things."

"But apart from that, do you instruct the miners themselves who are in your Union, as to their duties under the rules?"

"No."

"Would it be a part of your duty as secretary to see that they understood the business of the mine — under the law that they were mining under?"

"Oh no, it would be too difficult a task." [5]

Certainly the union had translators. It provided one at the inquest when a miner was questioned who was unable to speak or understand English. But they apparently had little or no interest in helping their foreign-speaking members understand mining rules and laws.

The questions, as well as Burke's answers, reveal all too clearly the reluctance of unions to assume even partial responsibility for the safety of their members. It was much easier to criticize mine management for its failure to deal with the problem. Consider the issue of the matches and smoking materials found with the dead miners. Had anyone told them that such behaviour was both illegal and dangerous? The question was neither asked nor answered at the inquest.

The Bellevue explosion happened on the evening of the ninth, but Hutchison at Hillcrest did not receive word of the disaster until the next morning, as he met with the pit boss and the two miners contracted to drive the rock tunnel. Brown, who had personal, painful experience with mine accidents, had been at Bellevue all night helping in the rescue and recovery efforts. When word came at about seven in the morning that rescuers had become trapped in the mine, Brown immediately ordered horses to be hitched up to a bobsleigh, which then raced a crew of Hillcrest men to the Bellevue mine just across the valley. It was Hutchison's first experience with a mining disaster, and he discovered the awful toll firsthand. The rescuers had already found the bodies and lined them up inside the mine by the time Hutchison and Brown got there. All Hutchison and Brown could do was help load the bodies into mine cars and then push the cars along the tracks to the mouth of the level at the slant. It was cold, wet work, and although coffee and sandwiches were served to rescuers at the mine office, Hutchison decided, when the work was complete, to walk back in the bitter December cold to get a hot meal at the Union Hotel in Hillcrest. It was less than a mile's journey, but the cold nearly killed him. Shortly after he left the Bellevue mine, his wet trousers had frozen solid, seriously hampering his ability to walk. As a result, he barely made it to the hotel. A rescue effort and a short winter walk had come close to making Hutchison another fatality and a side story to the deaths of thirty-one men in the dark tunnels of the Bellevue mine.

CHAPTER 6
THE DRIFT TOWARDS DARKNESS

Hutchison had received a nasty introduction to the dangers of coal mining with his assistance at Bellevue, but it caused only a minor interruption to his own development activities at the Hillcrest Collieries mine just across the valley. He oversaw the installation of new hoists: one at the entry to the rock tunnel, which was also the entry to the new no. 1 slant, and another at the no. 2 entry to replace the old, inadequate hoist built by Hill in 1905.

Development work continued into 1911, but it also brought a strike. On April 1, more than 7,000 miners, all members of the UMWA's District 18, walked out to press for higher wages. Hillcrest Collieries, as a new member of the Western Coal Mine Operators' Association, was struck along with all of the other members. General Manager Brown had attended his first meeting of the association on February 14, 1911. The Crow's Nest Pass Coal Company, which had withdrawn from the association in 1909, had re-applied and was re-accepted for membership on

February 22. Among those present at a March 2 meeting of the association were James Ashworth of the Crow's Nest Pass Coal Company and W.D.L. Hardie of the Diamond Coal Company, who would be elected mayor of Lethbridge in 1913. These were two of the big players in the coal fields of the Pass, and all of them knew each other. The operators met with the union in session after session in an effort to reach an agreement, but to no avail. The operators proposed that the whole matter should be submitted to a board of arbitration as a way to break the deadlock. The miners refused, and the strike ensued, long and bitter, until outside pressure came to bear upon it.

The smelters at Trail, Grand Forks, Nelson, Greenwood, and Boundary Falls in British Columbia were all affected by the nine-month-long strike. As well, the Government of Saskatchewan began to apply pressure out of concern about a possible coal shortage over the winter. Finally, the miners applied for activation of the Board of Investigation and Conciliation under the Lemieux Act. The operators, however, disputed the board's findings. Furthermore, in a letter to the president of District 18 composed by Lewis Stockett, the president of the operators' association, P.L. Naismith, the head of the CPR Natural Resources Department, and legal counsel Colin MacLeod, the operators demanded recognition of a fact that seemed to have been overlooked: "the report states that one of the facts disclosed by the investigation of the Board, was that probably two thirds of the mines in the association have operated, during the past two years, at a loss."[1] In other words, times were bad for the operators as well as for the miners, and the operators, therefore, rejected the board's findings, which were favourable to the union.

The difficulties presented by the strike for miners and their families are difficult to imagine. At Hillcrest, Alice Lote, daughter of a Hillcrest fire boss, had to work with her mother as domestic help. Barely fifteen, she scrubbed floors for a dollar per day, which she gave to her family to buy food.[2] It was similarly difficult for miners' families. The strike ended with a settlement on November 18, 1911, just in time for the re-opened mines to supply coal for the winter. The strike, with all of the hardships it imposed on miners and their families, resulted in a moderate wage increase. The union, however, conceded an open shop to the operators, as well as the insertion of a non-discrimination clause. The open shop meant that miners hired by the companies were not required to join or financially support the union. The non-discrimination clause meant that companies were not forced by union agreement to exclude Asians from mines.

At Hillcrest, management took full advantage of the downtime. The original mine office had been a large log structure situated on the north side of Drumm Creek, close to the mine entry. Two clerks, the accountant, and Hutchison not only worked in the drafty, poorly insulated building, but also lived here. This type of arrangement also existed in Western lumber camps of the day. On paydays, the miners would assemble in the main office to receive their cheques. Hutchison slept on a cot in a lean-to attached to the side of the log structure. A stove in the centre of the office kept the building heated, but the winter of 1910 was so bitterly cold that he and the two other men moved their cots out of their bedrooms and positioned them next to the stove while they slept. They didn't get much sleep, though, because the parts of them farthest away from the stove would still get cold, and one of them would have to rise in the night to put more coal in the heater. The miserable conditions would last until late in 1911. During the strike, management built and equipped a brick lamp house, an office for the fire bosses and mine manager, blacksmith and carpenter shops, car barns at the mine site, and new company buildings in town.

Among the more significant of the improvements was the new powerhouse, which would supply electricity not just to the mine operations (such as the new steel tipple, designed by Bill Hutchison, erected by a firm of American contractors and completed in 1912), but also to a new Sheldon sirocco electric fan, installed in the fall of 1912 to ventilate slant no. 1, and the water pumps at the bottom of the slants. The generator would also supply power for the town. By that fall, District Inspector of Mines Francis Aspinall reported after an inspection, that the Hillcrest mine employed 120 men and four horses on the morning shift, ninety men and two horses on the afternoon shift, and ten men and one horse on the night shift. They produced 750 tons of coal per day.

The noise of new construction also resounded in the town. By December 11, the excavation for the cellar of the new Cruick-shank/Burnett store was nearly complete. Above the store, the owners planned to build a large Masonic hall. The new lodge, which included a reception room, dining room, dressing room, and, importantly, a piano, would finally open on April 16, 1913. The owners would use the hall, they announced, for social occasions as well as for lodge work. The newspaper also reported that Frank Earp, who had built and operated a boarding house in Hillcrest, had turned the building and furniture over to Robert Petrie, who then held a dance and social for all of the young people in town. Hillcrest had begun to build a sense of community.

Also in 1912, the provincial government, in concert with mine operators, held the first course to train men in mine rescue on the Alberta side of the Crowsnest Pass. The Bellevue disaster had alerted Alberta mining officials, owners, and the union alike of the need for modern rescue apparatuses, as well as men who knew how to use them. Miners from Coleman, Blairmore, Bellevue, Frank, Hillcrest, Burmis, and Passburg all completed the course. Among those who attended were Hugh Hunter, a Hillcrest miner, Dan Briscoe, a fire boss at Hillcrest, and Francis Aspinall of Leth-bridge. Aspinall's attendance can be taken as an indication of how seriously the government mines official regarded the issue of mine safety.[3] Provincial officials equipped a mine rescue car with the latest rescue apparatus, which included two Draeger pulmotors and Fleuss oxygen tanks and masks for use in poisonous atmospheres. The theory was that the car would be used both to train miners in using the apparatus for rescue, and would also be rushed to mines where accidents had occurred and supply the equipment for rescue and recovery.

Aspinall had inspected the Hillcrest mine on July 24 and 25, 1912, and for the first time had reported significant quantities of gas in the mine. The new fan would apparently help solve the problem, but his report was a warning that the coal strata in the mine were now releasing significant amounts of explosive firedamp.

Through the rest of 1912 and into 1913, production continued to rise. This increased production and the extension of activity deeper under the rock had both contributed to the worrying amounts of firedamp noticed by Aspinall. By February 3, 1913, the output had risen to 1,000 tons per day, produced by 142 men and five horses on the morning shift, 138 men and five horses on the afternoon shift, and 10 men on the night shift. By then, Hutchison had overseen installation of the new fan, too.

In June, water became a problem, flooding the mine until mechanics could install electric pumps to pump it out. Thereafter the electric pumps removed about 250,000 gallons of water from the mine every day.

On June 24, 1913, Aspinall returned to the mine for another inspection. He said in his report that he noticed considerable dust in the mine, in particular at the face of the no. 1 south entry, and in the counter entry and rooms. He informed mine manager Quigley about his observations, who responded that no shots were allowed to be fired in the rooms off this entry. Aspinall replied that he thought it would be advisable to discontinue blasting shots in the no. 1 entry altogether. It was to be Aspinall's last inspection of the Hillcrest mine. On October 1, he left to work as an inspector of the coal fields at Edmonton.

The new district inspector, Andrew Scott, headed to Hillcrest on January 31, 1914, and found that the company had accelerated coal extraction to a daily output of 1,400 tons. In the darkness of the mine, 160 men worked on the day shift, 125 on the afternoon shift, and twenty on the night shift. The mine employed a total of 396 men above and below ground, and used 350 shielded, double-gauzed, self-lighting Wolf safety lamps. Ten leather-helmeted horses pulled the trips along the levels to the slants from where they were hauled to the surface by the two hoist engines. In his report, Scott said that he found the general condition of the mine and its ventilation to be good. He also found a problem.

During his inspection, Scott found a couple of dummy cartridges that were, apparently, to be used to tamp shots. The problem was that one was filled with fine coal dust, and the other, half with fine coal dust and half with clay. The use of coal dust was contrary to regulations, and also dangerous. At John Sterling's urging, Scott wrote a letter to Quigley, calling his attention to the importance of this violation of the Mines Act, and insisting that he strongly impress upon shot-lighters and examiners that they must comply with the law in this matter. Quigley wrote back that he "was surprised and felt vexed myself as it was the first time to my knowledge that I have known of anything but clay being used for over four years." Inquiries by Scott and Quigley determined that the dummies were made by a new man who could not speak English. His partner, an old-timer, did not know that the new man had put coal in the dummies. The matter was concluded with a reprimand to the miners, and instructions to the fire bosses that any miner who did it in the future was to be sent out of the mine to Quigley, who would deal with the matter personally.

While operations at the mine appeared to be going reasonably well, the population of the town of Hillcrest had risen, in concert with increased production, to about 1,000. The mining families included a diverse, if not integrated, ethnic mix of Italians, Ukrainians, Scots, Irish, English, Belgians, French, Americans, and Canadians.

Scott inspected the Hillcrest mine again on April 6, and reported that production was down because of the decreased domestic demand as warmer weather approached. One hundred and forty-five men on the day shift, eighty on the afternoon shift, and twenty-five on the night shift, along with seven horses, produced 1,000 tons of coal per day. Scott found the ventilation and timber in the no. 2 mine good, with no explosive gas present. He did not mention mine no. 1, but on May 18, the pit committee, made of up of three workmen appointed or elected by the UMWA local found gas in two rooms, but found that ventilation and general conditions were good. This would mitigate the dangers of firedamp, but only in the face of proper safety procedures.

On June 17 and 18 of 1914, management closed the Hillcrest mine for two days because demand for coal was down seasonally. But the fire bosses continued regular inspections. At 10:30 PM on June 18, fire boss Dan Briscoe exited the mine and reported that he could find only a small quantity of gas in the no. 3 south entry. He reported the amount of coal dust as "normal," with lots of moisture in the mine, which would mitigate the danger of dust. Dust and water combined is nothing but mud.

At 7 AM on June 19, when the whistle atop the engine house blew to call the miners to work, it also awakened Hillcrest storekeeper George Cruickshank. Cruickshank had come west in 1898 with the intention of heading to the Klondike, but heard so many tales of woe in Winnipeg from men who had been there that he decided not to go. Instead he arrived in Calgary in 1905, but by the following year had moved to Frank, in the Crowsnest Pass. Cruickshank worked for a storekeeper there for four years, where he formed a partnership with a friend, opened a grocery store in town, ran it for a year, and then together bought the McLean

Trading Company in Hillcrest.[4] They built a new store in Hillcrest in 1910. Cruickshank had married his wife Jean in 1909.

June 19 was a work day at the mine for 35-year-old fire boss William Adlam, who finished his shift at 7 AM. He had worked in mines for twenty-three years, fourteen in England and nine in Canada. At Hillcrest he walked the mine with a young assistant named Freddy White. Adlam claimed that he had entered the mine just before 4 AM, conducted his inspection, and exited at 6:20. He reported that during his inspection he found two new cave-ins and indications of gas in seven rooms, but no more than usual. He reported that he found enough gas to put out his light in six rooms, which he fenced off. Adlam's barometer, which read 29.4 inches, indicated that there was a larger volume of gas in the mine than normal. He found the most gas, enough to put a light out, five feet from the face in chute 43. As fire bosses always do at the end of their shift, Adlam posted his report so that all miners and management could see it before they went into the mine for the day. The fire boss' report is the safety bible for the shift. Tired from the night's work, Adlam then walked in the sunshine down the mine road to the town for breakfast.

Before each shift, the men lined up at the office of the mine's timekeeper, Robert Hood, who gave each man two numbered brass tags, or checks: a square one and a round one. Each man handed the square check, or lamp check, to the fire boss once his lamp had been examined and lit. On this day, fire bosses John Ironmonger and his cousin Sam Charleton did the inspecting, looking for flaws that might make a lamp dangerous. The fire boss then put the square check on a nail in the lamp house; each particular nail indicated what time the miner started work, and where he would be in the mine. During the shift change, Ironmonger and Charleton apparently found the lamps to be in good shape. Ironmonger then entered the mine with superintendent Quigley and then the two men separated. Two hundred and twenty-eight men entered the mine at 7 AM, including the boys who operated the hand fans to ventilate the raises. Inside, Ironmonger found no traces of gas or cave-ins. Between 8 and 9 AM, he fired five shots to loosen coal at the face for the miners. From his location off the no. 2 slope, miner George Wilde was helping build stoppings to direct the flow of air when he heard the shots. "No unusual disturbance followed them," he said later. "I saw nothing unusual prior to the accident in the mine. The air was good and there was no gas."

At 9 AM, seven more men entered the mine through the rock tunnel at the No. 1 entry. There were now 235 men in the Hillcrest workings. Two other men who took checks did not enter the mine, but worked outside in the yard. At least three men had been scheduled to work that morning, but did not. Hillcrest fire boss Dan Kyle, a man of extraordinary good luck, was one of them. After deciding to leave Britain and come to Canada, Kyle booked passage on the maiden voyage of a large passenger ship, but cancelled the booking when a good friend, also bound for Canada, was unable to get a room on the same ship. Instead, they travelled together on a smaller liner in early 1912. The ship on which he had first booked passage, the *Titanic*, never reached its destination, and now rests at the bottom of the Atlantic. On this day, Kyle had been told to show up for the afternoon shift. While others worked, he stayed home.[5]

Julien Lefort was scheduled to work, but did not feel well. François Labonne took his place. Labonne's sister's husband and his wife's two brothers all walked up the hill to work in the mine.

Another man who had been scheduled to work, Thomas Price, a native of Lancashire, England, had changed shifts so that he could go to the Hillcrest train station to meet one of his brothers, who was scheduled to pass through on the early morning train.[6] The replacements for Kyle and Price entered the mine on the morning of the 19th. George Hicken, a twenty-seven-year-old native of the small village of Norton Canes in Britain, had changed shifts. Hicken was the cousin of fire boss

Sam Charlton, and of the three Ironmonger brothers: John, a fire boss, Charles, a roper rider, and Sam, a timber boss. All were scheduled to work in the mine that morning. Hicken walked up the hill to the mine with his cousins, leaving his wife Lily at home with their infant son George. They had been married just two years. Another man who was not scheduled to work, but who had changed shifts, was Alphonse Heusdens, a Belgian immigrant who also left a wife and infant son at home. He walked to the mine with the others.

At the wash house, the men changed into their work clothes and stowed their street clothes in lockers. From there, they walked the short distance to the lamp house, where they lined up for their brass checks and lamps as the fire boss made sure the lamps worked properly. That done, the men walked, lunchpails hung on shoulders, over the granular coal that covered the ground, along the mine tracks, past the no. 1 hoist house, through the wooden timbers at the entry, and into the cool, electric, incandescent bulb-lit dimness of the rock tunnel that had been blasted out four years earlier. The fan hummed on the hillside. The voices of the men echoed off the cold, damp rock, and footsteps crunched on the floor, then faded as the men shrank into the dim distance of the tunnel.

Two men who walked into the Hillcrest mine on that day — Charles Elick and John Hood — held the distinction of having also been among the seventeen miners who found themselves buried underground during the Frank slide in 1903.[7] Fighting off panic, they had managed to dig their way to the surface from the top of a raise only to discover that a rock slide had smashed to splinters, swept away, and buried all of the mine's surface buildings under millions of tons of limestone boulders, along with part of the town, including some of the miners' cottages.

At the Hillcrest mine, on the morning of June 19, 1914, assistant engineer Alex May had just entered the engine house where his father, Jack, maintained the boilers. It had taken him ten minutes to walk back from the no. 1 fan house. It was 9:20 AM. In the town, women went about their household chores and children sat at their desks in school. The morning brought sunshine and a mild temperature, a promising spring day in the Pass, the kind of day that makes one feel good to be alive.

Three miners before their "Riverside Villa" at Hillcrest: a shack of wood and tarpaper with a window above the door. A laundry tub hangs on the wall.
(Belle Kovach)

Hillcrest miners at the no. 1 hoist house in 1912. George Frolick, far right. Note the round brass mine checks attached to the hats of some of the men, the lunch pails, the safety lamps, and the drill, far left, used to bore holes into the coal for the insertion of explosives. (Vernon Frolick)

X

Charles Plummer Hill in 1919: born in Seaford, Delaware, he becomes a naturalized Canadian in 1904 and marries Enid Mary McLean, the daughter of a Charlottetown miner who had come west. In 1905, he established the Hill Crest Coal and Coke Company. (Glenbow Archives)

Bellevue miners at shift change c. 1910. The fire boss (centre) carries the coil of electric wire used to detonate explosives that loosened coal at the face, where it was extracted by miners. (postcard in author's collection)

Heavy Draeger helmets, like these at Coal Creek, B.C., were rushed to the scene of the Bellevue explosion in 1910. (Glenbow Archives)

(OPPOSITE) The Hosmer rescue team with Draeger breathing apparatus and a pulmotor, 1910. The hero of the Bellevue disaster, Fred Alderson, kneels, left. (Provincial Archives of Alberta)

Hillcrest Collieries surface plant, facing west. Partly obscured by bushes is the wash house, centre left, and just to the right is the engine house. Between the two buildings, in the distance, the lamp house is visible. Turtle Mountain rises to the right. (Crowsnest Museum)

Hillcrest miners, survivors, and victims, ca. 1911. Identified, L to R: A. Cameron; Alex May; sixth from left on dinky, Torrie Hood; seventh from left, mine manager James Quigley (killed); far right, John Hood (killed). (Glenbow Archives)

The Hillcrest no. 1 entry to the rock tunnel. The wires carry telephone signals and electricity for the lights and pumps. (Vernon Frolick)

(CLOCKWISE) Alphonse Heusdens agrees to work in place of his friend, Joe Labourier, so that he can meet his wife, scheduled to arrive on the train that morning from Belgium. Heusdens would leave behind a wife and infant son, Peter. (Peter Heusdens); Miner George Hicken, cousin of fire boss Sam Charlton, is not scheduled to work, but changes shifts at the request of another man. He would leave behind a wife and son. (Louise Wells); George Frolick in 1913. (Vernon Frolick); Miner John Hood is killed by carbon monoxide, known to miners as afterdamp. (Glenbow Archives)

The first Blairmore rescue team, wearing Fleuss breathing apparatuses. (Glenbow Archives)

Members of the recovery team push a mine car containing a body from the no. 1 entry past the damaged no. 1 hoist house. (Belle Kovach)

Miners and a RNWMP officer, his Stetson visible, centre, unload a body from a mine car at the engine house. From here men carry the remains down a set of stairs to the wash house where they are, if possible, identified, stripped, washed, and wrapped in fabric from George Cruickshank's store. (Glenbow Archives)

June 19, 1914: worried friends and
relatives await word at Hillcrest Collieries
entry no. 2. A makeshift hospital is set
up under tarps next to the hoist house.
(Provincial Archives of Alberta)

Perhaps no photograph taken at the Hillcrest disaster represents the terrible loss to the families of the 189 men killed as this one, published in *The Canadian Courier: The National Weekly* on July 4, 1914: a pregnant mother with her little boy, handkerchief in hand, as she awaits word of her husband. (Toronto Reference Library)

The day of the funeral started out pleasant enough, but turned cold, with skin-numbing particles of ice and dust driven by a harsh wind. (Glenbow Archives)

The funeral procession to the Hillcrest cemetery. On the far left is the Cruickshank store, in the centre the Union Hotel, and directly above the hotel, the tipple. (Vernon Frolick)

Compressed air locomotives at Hosmer. (Provincial Archives of Alberta)

The Hosmer entry as it looks today. (Steve Hanon)

Mourners at the Hillcrest cemetery gather beside the mass graves.
(Provincial Archives of Alberta)

(CLOCKWISE) George Cruickshank in his Masonic regalia: the trusted Hillcrest merchant helps identify bodies at the wash house. (Belle Kovach); Elizabeth Murray in 1938: a woman of extraordinary resolve and courage, who lost her husband and three sons in the disaster, she worked as a cleaning woman to raise ten children through the Great Depression. (Barbara Allan); Lily Hicken with her son George, shortly after the disaster. Widowed and broken-hearted, she died at age thirty-two. (Louise Wells); William (Bill) Hutchison in uniform, circa First World War. Hutchison helped restart the no. 2 fan, then became the first rescuer to enter the exploded mine. No man contributed more to the success of Hillcrest Collieries. (Jim Hutchison)

Hillcrest miners leave the mine after their last shift, before it was closed for good in 1939. (Peter Heusdens)

Cortlandt Starnes, commissioner of the RCMP, a position he held from 1922–1931. In 1914, he supervised RNWMP officers at the Hillcrest disaster. (Glenbow Archives)

Hillcrest Collieries after closure, in 1940. Anything salvageable, such as boilers, generators, fans, steel rails, and mine car wheels, was removed and hauled away. Sniffy the dog surveys the scene. The wall of the car repair shop, right, leans precariously. Of the tracks, only the wooden ties remain. (Rick Petrone)

The ruins of the Hillcrest Collieries engine house, where steam engines and generators provided power for the mine and the town. (Steve Hanon)

A blueprint of the Hillcrest mine in 1914, used at the Commission of Inquiry. Within the shaded area is the origin of the 1914 explosion. The rectangles and squares represent pillars. Where they are shaded, the pillars have been extracted. (Frank Slide Interpretive Centre)

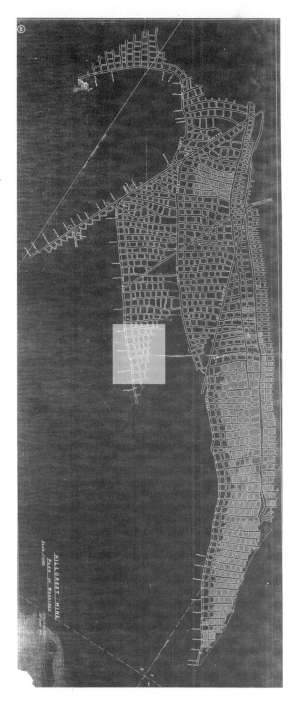

CHAPTER 7
WITHOUT AIR TO BREATHE

In the darkness and confusion, amidst the screams of pain and pleas for salvation, you hear the voices of calm from experienced miners. They try to offer reassurance and quell the panic. Some suggest that you wait for rescue teams who, they say, will arrive quickly. Others advise trying to escape on your own, using the rails to guide you in the right direction. This makes sense to you, and you decide to flee. Your horse lurches to its feet, and you grab its mane, half dragging yourself, half walking along. You shout for others to follow, and some do. You hear the voice of Budzynsky, a countryman. In his agony, he begs you to kill him. Both of his legs are crushed. Barely alive yourself, you cling to the horse as it struggles on, instinctively, in the darkness. It sways and makes sounds that animals do not make. You pass men you cannot see in the blackness, men unable to move and who beg that you deliver last messages to their loved ones. Each member of your small group of survivors reaches deep into himself for strength to fight against the whisper in his ears to give up, to sit down, to rest, to sleep. Somehow, you find the strength to ignore it, and whatever raw part of yourself you tap into gets you to *your first goal: the junction of the level and the slant. But finally the horse fails in its struggle to breathe, and collapses. Foul air and heat force you to crawl on alone, over shattered timbers and debris. You reach a raise and feel cool air wash over you. You cry out to your companions who have fallen behind. Some manage to crawl to you, and then you, with renewed strength, crawl back to drag out the three men unable to move on their own. You rest for a few minutes, gathering strength in the life-giving air, then crawl on in the pitch black. You see, at last, light ahead, and your hopes rise. But you realize that the light comes not from the sky, but from burning timbers and shoring. You crawl closer to the flames, where you expect to die. Your heart rises in your throat when you see that only one side of the tunnel burns. You hug the far side, and drag your uncooperative body past the danger. You continue your slow, agonizing crawl until you see only the tiniest point of light. The mine exit! That's it! But for you, the struggle is over. You and your companions, overcome by afterdamp and exhaustion, sink into unconsciousness.*[1]

The stories of escape from the no. 1 north level off the no. 2 slant that day display a repetitive, dull similarity, but beneath those stories lies an adrenaline-charged flight from death by each man who got out alive. As George Wilde worked on stoppings to close off passages and redirect airflow, he heard a noise that he described later as similar to that of a blown-out shot, then an unusual sort of buzzing sound. In a place where anything unusual may well be dangerous, he decided to investigate, and with his safety lamp climbed down to the level. There Wilde was met by smoke so dense that he could see absolutely nothing. Something had obviously gone very wrong. The hot, smoky air made breathing difficult, and in the absence of ventilation, he assumed correctly that the fan had stopped working. Wilde then began to move, his steps more and more urgent. His heart pounded as he realized, with growing horror, that he might not escape. A fast walk turned to a run towards the no. 2 slant, but before he

reached it, a small gift came to him in the form of returned ventilation. It brought cooler air and less smoke, which made his run
easier. When he reached the slant, the seriousness of the situation became all too apparent. Wilde found himself confronted
by a debris field consisting of broken mine timbers and other
wreckage strewn about on the slant, the only escape route to the
surface. He began to step around and over the wreckage, but in
his effort to hurry, he lost his footing and fell painfully. His lamp
slammed hard to the ground and went out, but as he picked himself up in the darkness, suddenly he could see the light of other
lamps carried by other men hurrying up the slant in a flight from
deadly gases. Together, they hastened up the slope to the surface.

William Guthro worked a short distance inside the no. 1 north
entry when it happened. It took him at least ten minutes to cover
the considerable distance from his workplace to the slant. Guthro,
well known throughout the Crowsnest Pass, had been in two previous disasters. But now, on his way up the slant, as he followed
the mine car rails, his boot became stuck in a frog (the crossing
point of two rails) in the track.[2] Panicking, he struggled without
success to free the boot, then pulled a pocket knife from his trousers and began cutting at the laces and leather. Finally, his boot
in ruins, Guthro yanked his foot free. Now with only one boot,
he limped on as fast as his sock-clad foot would allow over the
sharp rocks and pieces of coal, upwards to the entry, and into the
fresh air and sunshine.

Pete Dujay felt the explosion, but like many others, he heard
no sound. He hurried upwards through thick smoke, also fell over
obstructions on the slant, and came across two horses and three
men, all dead. He got out at about the same time as Guthro, but
said that he was nearly finished off by the smoke and gas.

Jack Maddison was at work about 1,500 feet from the mouth
of no. 2 when he heard a loud report. "I made a rush for the chute
and up the slope. It was a long walk and I felt very keenly the
effect of the gas. At times I felt inclined to lie down, but I did not

yield to the inclination, and finally walked to safety. On my way out I passed dead bodies on either side of me."[3] Among the dead men that he knew were two track layers, Rod Wallace and Billy Neath. He and another fellow miner, L.J. Lott, passed Fred Muirhouse, also dead in the rubble, and came upon William Guthro, who was limping up the slant after having cut off his boot.

Joseph Atkinson, also in slant no. 2,

> "did not hear the report of the explosion. It was just as if I had suddenly gone deaf or as if two four inch nails had been driven in my ears. That is how it felt. I was bowled over by the shock but scrambled to my feet. Almost instantly thick black smoke began to come around the slant [level] from the direction of slant No. 1. I was almost overcome by the shock and the smoke but started to run towards the mouth of No. 2. There were several workers with me and they did the same as I did. A short distance ahead we came on Billie Neal [Neath]. He was lying on the ground overcome. We tried to lift him and carry him along with us but by this time we were too weak with the gases. We carried him for a short distance dragging him over a dead horse that had been killed in his tracks and then we had to drop him. I shouted, 'Come on boys, we're all in anyway,' then we came to the afterdamp, a solid wall that drove us back, nearly suffocated us...."[4]

In a story published in the *Morning Albertan* on June 22, Atkinson adds that they were forced back to a pool of water. They tried to breathe through water-soaked sleeves, but the effort failed, and they lost consciousness.

Whereas Atkinson did not hear the explosion, his partner, Herbert Yeadon, did. According to Yeadon's account in the *Coleman Bulletin,* he was working with five or six others about a quarter of a mile in from the mouth of no. 2. "It sounded very much like the discharge of a big gun on a battleship. Men were running towards us, and we could see that there had been a couple of men killed. We all turned and ran up the drift of No. 2, but were suddenly driven back by black damp ahead of us. We ran back and lay down near a pool of water … Here we must have been overcome by the fumes, for I do not remember anything else till we came to outside the mine." Physicians put Yeadon on the Pulmotor for forty-five minutes. Upon resuscitation, he immediately assisted in rescue efforts.

Next to Atkinson and Yeadon, Alphonse Heusdens and his partner, 37-year-old Gustav Franz, were near the junction with the no. 2 slant at the time of the blast, likely handling coal cars. Both were killed, victims of afterdamp. Franz's body was found at the first parting of no. 1 north, and Heusdens' was found nearby. They were the only men who worked in no. 1 north who perished.

In the next chute over, August Kovach, a man who had built his family a house at Passburg, was working on level 1 north with a friend, Mike Semancik, when they realized they had a serious problem. Like the others in no. 1 north, they headed towards the slant as fast as their legs and lungs would allow, but Semancik stopped, overcome by the smoke and afterdamp. Gasping, he indicated to Kovach that he couldn't go on. Kovach considered the situation for only a moment, and rather than flee alone, leaving his partner and friend behind, he grabbed Semancik and dragged him to safety.

Fire boss Ironmonger later said, "The first intimation I had of anything wrong was a concussion. There was no noise. I started to go out and met smoke coming in the entry. There was no heat." He and other nearby men rushed together towards the slant.

Malcolm Link worked with Charles Jones and another man at chute 15, nearer the mouth of the drift than Yeadon. For them, the explosion sounded like a massive gun. Jones was knocked flat. They ducked low and made for the entry. By keeping close to the ground, and at times crawling on their hands and knees, they managed to get through the black damp (an atmosphere very low in oxygen) to safety. Bodies were piled in heaps, Link told reporters, along with pieces of twisted tracks, cars, horses, and mine timbers.

Arnold Varley told a reporter, "I just heard the report, and then I rushed to safety. There were a number of others around me, and I can remember stumbling over a dead horse on the way out." Arnold and his brother Herman both escaped.

George Weatherington worked just a few yards from Jack Sands, Billie Moore, and Pete Hawkins. He escaped, while they perished. Their workplace: no. 2 south. His escape, like that of George Frolick and the others who fled the same workplace, is a testament to luck, determination, or both.

On the surface, the blast forces a high-pressure column of air from the no. 1 rock tunnel with so much force that it demolishes part of the eight-inch concrete wall of the no. 1 hoist house facing the mine, and blows off its roof. Pieces of concrete tumble to the ground. The explosion lifts up 19-year-old rope rider Charles Ironmonger, brother of the fire boss, and hurls him like a shot from a cannon into the side of the hoist house. Men find him on the ground, barely alive, but he dies shortly after being admitted to hospital.

The chief engineer, 39-year-old William Hutchison, is with his younger brother David, a civil engineer and mine surveyor, on the no. 3 slope, an abandoned part of the mine close to the no. 2 entry. He is conducting a survey when the explosion occurs. He has just set up his survey instrument when they are both enveloped in a tremendous wind and can barely stay on their feet. The wind is followed about thirty seconds later by brown smoke. William hears nothing, but realizes immediately that the mine has exploded. He turns to David and says, "The mine is up." Both men immediately make for the surface as the smoke gets thicker. William detects no gas. Within a minute he reaches the mouth of the no. 2 slant, which is also belching smoke.[5] At the no. 2 hoist house he meets general manager Brown, who was within 60 or 70 feet of the opening of no. 2 mine when the smoke issued. Brown is covered in black dust and barely recognizable when he meets Hutchison. Brown's own brother, William the pump machinist, is somewhere in the mine.

Hutchison tells Brown that he will enter the no. 2 slant with his lamp, but Brown sees that the force of the explosion has stopped that mine's fan. Brown asks Hutchison to see if he can get the fan started again, but with the direction reversed, then he and David Hutchison run toward the other fan at no. 1. Starting the fan in reverse will draw a current of fresh air directly down the no. 2 slope, into the faces of the men trying to escape. Brown and Hutchison know that the survival of the men below depends on fresh air.

The man who looks after the no. 2 hoist and fan, Thomas Brown (no relation to the manager), is temporarily blinded by the dust and dirt. He clears his eyes as best he can, whereupon he and Hutchison discover that the fan is not damaged, just stopped. Hutchison gives the fan a pull with his hands, and with the force of the steam behind it, it starts up. According to Hutchison, only three to four minutes pass from the time it stops until they start it again. Now it begins sending fresh air down the slope into the no. 2 mine.

The outside assistant engineer, Alex May, has just entered the engine house after inspecting the no. 1 fan. After the Lille mine had closed in 1912, Alex and his parents moved to Hillcrest and into the house that Hill had vacated when he sold the mine.[6] Alex's twin brother Albert works as the power house stationary steam engineer at Bellevue. Another brother, Jack, is on the train crew that hauls railway cars from the tipple to the Hillcrest station yard. Now, near the no. 1 entry, and puzzled by a muffled sound and the sensation of a slight tremor, May rushes from the engine house to see what has happened. He is stunned to see black smoke and pieces of rock shoot from the fan fifty feet into the air. He realizes that if he had been at the fan, he likely would have been killed. His father, fifty-five-year-old Dan May, shift steam engineer at the engine house, blows the large steam whistle on the power house roof three times — the signal that an accident has happened. The no. 1 fan, damaged in the blast, has stopped. This means that foul air, particularly the poisonous carbon monoxide produced in the explosion, will not be evacuated from the recesses of the mine where men now are desperately trying to escape.

The two dinky engineers, George Porteus and Torrie Hood, both have brothers inside the mine. Porteus and Hood are in the mine yard with the dinky engine, waiting for a load of coal, when the mine explodes.

Twenty-one-year-old Jonathan Penn is employed as a coal sampler for the CPR at Hillcrest. At the moment of the blast he is working near a locomotive at the tipple. He hears nothing, perhaps because of the noise around him, but seconds later becomes aware that something is wrong when he notices a heavy cloud of dust and smoke belching from the mine at the no. 1 fan and entry. His father, at sixty-five the oldest man in the mine, and his younger brother are both inside.[7]

The steam whistle alerts the residents of Hillcrest. The miners' wives recognize the urgent signal that now repeats over and over. They step outside to look up toward the mine, and turn pale as they see the tall column of dark brown smoke rise into the air. They grab their children, and make their way to the wagon road that angles steeply up the ridge to the mine. Soon, a frantic stream of women, children, friends, and other miners clog the road, some on foot, others on horse. Among those who rush up are RNWMP officers on horseback, fire boss William Adlam, Coleman chief of police Ford, and Dr. Allan Ross of Coleman. Ford, a former miner himself, escaped death in the Bellevue disaster in 1910 along with his then-partner, Arnold Varley, who has already escaped the Hillcrest mine by the time Ford gets there.

At the mine entries, the crowd, which soon numbers in the hundreds, presses forward. Some men and women, including miners, hysterical with anxiety for sons, brothers, and husbands, attempt to rush into the mine, desperate to do anything besides wait. Ford and Constable Paynter of the Bellevue detachment shout at them to stay back, but the crowd surges forward anyway. The policemen are suddenly aware that if they are going to stop anyone from entering the mine, they will have to take serious action. They cannot use their firearms, which remain holstered. Instead they are forced to use axe handles.

Shortly after Ross and Ford arrive, Dr. Connolly and most of Coleman's nurses rush to the scene, where some of them erect a makeshift hospital tent at one end of the no. 2 hoist house.

Across the valley at Bellevue, someone tells miner Harry White about the explosion. He looks across the valley towards Hillcrest, and sees a large plume of dense, dark brown smoke in the air above the mine. White had been a fire boss at Hillcrest from 1908–1912, so he knows the men who work there. More importantly, he knows the layout of the mine itself. As well, the experienced miner and fire boss had briefly worked at Hillcrest in November. He finds a horse, mounts it, and gallops to Hillcrest, worried about his brother, fire boss Sam Charleton, and his brother-in-law, Leonard Clarke, both of whom were scheduled to work that morning. He is unaware that another brother-in-law, George Hicken, is also in the mine. By the time he arrives at about 9:45 AM, the smoke has dissipated.

Another man in Bellevue notices smoke in the sky over Hillcrest. William Grafton is on the steps of the Southern Hotel when he sees it. When he rings up the Hillcrest exchange, he's told that nothing is wrong, but he suggests they make further inquiries: Ten minutes later, the Hillcrest engine house whistle blows its signal to report a serious accident.

Hutchison, meanwhile, has gotten the no. 2 fan going, and immediately steps into the mine with a safety lamp, but without a breathing apparatus. With him goes James Gorton, the secretary of the UMWA local, who had been at the Hillcrest Hotel when he saw smoke issue from the no. 1 fan drift. It is difficult to characterize the actions of these men as anything other than heroic. While others struggle to flee the mine, Hutchison and Gorton plunge into it, towards the danger, like a good soldier hurrying towards the sound of gunfire. With their lamps, they penetrate down the smoke-filled slant for about 400 feet before they meet anyone coming out. The men they encounter gasp for air after running uphill through the heavy smoke. Among the group of men are Jack Ironmonger, the fire boss, and Hugh Stevenson, a track layer. According to Hutchison, it might have been between five and seven minutes from the time he first noticed that something was wrong to the time he met the men running up the slant, but in reality it was likely at least double that.

Some of the men, inspired by Hutchison and Gorton's example, rest briefly on the surface to get their breath back, then try to re-enter the mine. But the thick smoke makes it impossible to enter, so they move lower, to a little below the no. 1 south. That is where they see the first dead victims. Timbers have fatally crushed two men, William Neath and Rod Wallace. The group continues on in a search for survivors. At the out-bye parting to the no. 1 north, Hutchison finds twelve loaded mine cars, two more dead men and a dead horse. One of the men, Gus Franz, lies face-down in water beside the horse. The other is crushed under one of the cars. A lamp burns nearby. Then Hutchison hears the sound of men breathing hard. They find a group of men unconscious. He helps carry three of them 100 feet or so up the slope, where quite a few men are now gathered, some from the outside. This was likely the group that included George Frolick.

Speed is now crucial to the saving of lives. Hutchison, Gorton, and another man head lower on the slant. They find no men, only a horse still alive, but they help it get to its feet and remove the harness, then walk it partway out. The horse finds its own way out and is the only one of its kind to survive the disaster.

Mechanics, meanwhile, have been working at the no. 1 fan on the shelf above the no. 1 entry in an effort to get it restarted. The fan is a Sheldon Sirocco wheel with a diameter of 9.5 feet. When it runs, driven by its 100 horsepower motor, it produces 90,000 cubic feet of air per minute. Finally, at about 10 AM, the damage repaired, they restart the fan, which immediately begins removing gas from the mine. It helps, but the explosion has blown away stoppings and brattices that control air flow in the mine. Until this is repaired, much of the gas and smoke will stay in the mine and interfere with rescue and recovery efforts.

Alberta Mine Rescue Car No. 1 sits on a siding three miles away at Blairmore where the car superintendent, Henry James, trains miners in rescue technique. James receives word of the explosion by telephone at 10 AM, and is asked to rush the rescue car, a modified rail passenger car, to Hillcrest at once. James immediately calls the CPR to ask for a locomotive to haul the car to Hillcrest. The CPR moves as fast as any railway can, but the process takes valuable time. A locomotive has to be found, assigned, moved to the car, hitched, pulled to the Hillcrest CPR station, then either switched again onto the Hillcrest railway tracks and pulled up to the mine, or disconnected from the CPR locomotive at the station and re-hitched to the Hillcrest locomotive to be pulled to the bottom of the tipple, the closest a railway car could get to the mine. The car does not arrive at Hillcrest until about 11 AM, and even then it can get no closer than the Hillcrest tipple. The gear, which includes Fleuss rescue breathing apparatuses, the Draeger Pulmotor resuscitation apparatus, and battery-powered electric lights, has to be removed from the rescue car, loaded into a horse-drawn wagon, and hauled from the tipple up a steep, rough wagon road to the entries above. The rescue car carries eleven sets of Fleuss breathing apparatuses known as "Protos," which, when freshly charged, will last for two hours, and six one-hour apparatuses known as "Salvators." Five sets of the two-hour apparatus are assembled and ready to go when the car arrives at Hillcrest, except for one small detail: instead of the lump soda used to charge the apparatus, stick soda is carried up to the mine with the units. The units do not work effectively without it. These five sets, including leather heat helmets, are taken up to the entry at slant no. 2. The rechargeable battery-operated lights, which had just been put onto the market, would prove useless.

Early in 1911, in the wake of the Bellevue disaster, the Alberta government had moved quickly to provide rescue equipment and to train men at coal mines around the province. A trained rescue team had proven its worth at Bellevue, and Alberta officials and mine operators could no longer dismiss such teams as impractical. The Alberta government and mine operators agreed to share the cost of these new facilities. Initially, the organizers had rented three buildings at Blairmore and equipped them as a temporary rescue station. They conducted the training in a smoke chamber, around the station, and in the Blairmore mine.[8]

When CPR vice-president George Bury receives word of the explosion, he rushes to Hillcrest to find out if he can help. He also alerts CPR president Shaughnessy of the explosion. He sends a total of eight telegrams to Shaughnessy in Montreal that day with updated details and total deaths. Bury informs Shaughnessy that three special trains had been sent to Hillcrest from Calgary, Lethbridge, and Fernie. Later in the day he says that reports indicate 230 miners were trapped, but adds that it is very difficult to get accurate information. His last telegram at the end of the day informs Shaughnessy that the hope that anyone else will be found alive has expired, and that small fires were burning that, if not extinguished, would ruin the mine. "There is very little excitement at the present. Women and children facing the situation with stoical fortitude which is most touching."

Alberta's Chief Inspector of Mines John Stirling relaxes in a coach on a train heading north from MacLeod to Calgary when he's approached by the conductor at about 10:30 AM and told that the Hillcrest mine has blown up. Sickened, Stirling changes trains at Granum, and immediately returns to MacLeod to get a train up to Hillcrest. The CPR passenger train is loaded with mine rescue teams, their gear, and doctors and nurses sent from Lethbridge and Coalhurst to assist the efforts at the devastated mine.

Stirling, a Scot born in 1875 at Newbottle, had been college-educated at Edinburgh. In 1898, he was appointed manager of the mines operated by the Lothian Coal Company at Rosewell, Scotland, and worked in that position until coming to Canada in 1908. For a year he managed Edmonton's Rosedale Coal Company. At the end of 1909 he was appointed an Alberta district inspector of mines, and then, in 1911, was quickly promoted to the province's chief inspector of mines. Devoted to coal mine safety throughout his career, and with an excellent reputation among miners and mine operators alike, now, at the age of thirty-nine, Stirling is about to plunge into a major crisis.[9]

At around the same time, the RNWMP commander at Pincher Creek, Inspector Christen Junget, receives a telephone call informing him of the explosion. Born June 13, 1876 in Denmark, Junget had served for five years with the Danish army, where he had attained the rank of lieutenant, but decided to immigrate to Canada. He joined the RNWMP (then simply called the NWMP) in Regina in 1899 as a special constable, and became a regular constable the same year. Junget spent more than thirteen years at Yorkton, Saskatchewan. While serving there as a staff sergeant, Junget had been sent to New Scotland Yard in London for two weeks to study the new police science of fingerprinting. He was the first officer in the RNWMP to do so. Junget too, now faces a crisis at Hillcrest.

Upon his arrival at the mine, Junget discovers that Corporal Frederick Mead has taken the first steps in handling the crowd and in the removal of the dead from the mine. Mead had joined the RNWMP as a constable at Regina in 1910, and had been immediately posted to Fort MacLeod. On November 11 of that year, he had been promoted to corporal.

Before leaving for Hillcrest, Junget had called Superintendent Cortlandt Starnes, commander of "D" Division at MacLeod, who in turn wired the RNWMP commissioner in Regina. Starnes gave a bare outline of the activity at the mine, and, frustrated that he cannot leave his post and head to Hillcrest himself, complained about the shortage of men at MacLeod. He sent Constable D. Wilson on the first train to the mine; by the end of the day, a total of eight RNWMP are at Hillcrest.

Born in Montreal in 1864, Starnes had had, to that point, a most extraordinary career. He attended infantry school in St. Jean, but in 1885, when an opportunity presented itself to take a paid vacation in the West as a lieutenant in the Canadian Militia, he grabbed it, only to find himself in the midst of the North-West Rebellion. As adjutant of the 65th Regiment, he marched with Inspector Bowen Perry from Calgary to Edmonton, where anxious residents feared an Indian attack. Within a year Starnes had joined the Mounted Police, and was appointed an inspector in March 1886. Starnes was short but robust, with broad shoulders, and known to the Indians as "Little Bull." During the Yukon gold rush he distinguished himself as an inspector with "B" Division of the RNWMP at Dawson alongside Sam Steele. Aside from his experience, Starnes' value lay in his tact, as well as his calm and unruffled manner, even in desperate circumstances.[10]

Hillcrest fire boss Dan Briscoe is at home when the explosion occurs, and is alerted by the engine house's steam whistle blowing its three-blast signal over and over. He rushes to the mine to see a crowd of people milling about. Briscoe gets a lamp, and is told by Brown, who has rushed from the no. 2 entry to the no. 1 fan, to go into the rock tunnel and see how far in he can get. No one has emerged from the rock tunnel and fear has begun to rise in the throats of those assembled. Briscoe enters at about 10 AM, but is only able to get to the end of the rock tunnel at the top of the slant before he is stopped by heavy smoke. He rushes back to tell Brown, who sends him back to the no. 2 entry, where the superintendent of the rescue car has been preparing the apparatus. Briscoe, who in 1912 was one of the men to take the first course in mine rescue on the Alberta side of the Crowsnest Pass, leads the first rescue team to wear apparatuses into the mine, down the no. 2 slant. His rescue team meets Hutchison, Gorton, and the others who were unable to enter the No. 1 north level because of dense smoke. They carefully enter the level and search for forty-five minutes, but find no bodies. While Briscoe and his team, with apparatuses, search no. 1 north, Hutchison acts as a traffic manager for rescue teams, and stays at the intersection of the slant and the no. 1 north level. As they wait, other groups of men try to enter no. 1 north, but Hutchison instructs them to go deeper down the slant where they are needed to hunt for survivors. Gorton and the others continue down on Hutchison's instructions, as does a second group of rescuers, also without apparatuses. When the men with breathing apparatuses finish their examination, Hutchison presses on with two men to the no. 2 south level in an effort to get to the face. They pass rescuers that he had sent down earlier with victims, some dead, some alive. Pulmotors are used in an effort to revive some of the victims, with mixed results.

In no. 2 south, Hutchison and his companions find the worst devastation of the explosion: every man and horse had been killed instantaneously, their bodies strewn about by the force of the blast. All of the support timbers are blown down. Timber is even blown out of the level and onto the slant. They are unable to reach the face because of gas. Instead of smoke, so evident in no. 1 north, no. 2 south is threatened by afterdamp. Hutchison and his two companions flee the gas with unsteady legs. At the slant, they find that the hoist is in operation again. Miners help them onto a timber truck, which is then pulled to the surface. As Hutchison slides from the truck, a doctor welcomes them with a bottle of brandy.

The first organized rescue party without oxygen includes Charles Woolatt and George Vicars. Woolatt tells a reporter with the *Edmonton Bulletin* that "[t]he mine was still heavily impregnated with gas, and it was with difficulty that this gang worked its way to the pillars in which the men had been working. Three of the rescue gang were overcome with gas, and had to be carried out. When we reached the death trap we saw bodies strewn in every direction. We brought out Joe Atchison [sic] with every appearance of being dead, his limp form bearing no evidence of life, but the doctors and miners got at him, and for three hours worked over him with artificial respiration methods. Finally, a slight flush appeared … on his cheeks. The young lad fought against them, and seemed in his dazed state determined to die, but after three hours he recovered, and was soon able to assist in the rescue work [ellipsis mine]." Atkinson is the first man recovered from the mine unconscious after the explosion.

While mine rescue car superintendent James has been working at the no. 2 entry, a number of men, anxious to enter the no. 1 mine, take it upon themselves to enter the mine rescue car parked below the tipple. In his absence at the no. 2 mine, James has put D.H. Hyslop of Coleman in charge. The miners remove the disassembled Fleuss units and rush them up to the mine, but amidst the chaos, they fumble with connections they do not understand. Hurried trips are made back and forth to the rescue car for information and help, where Hyslop sits in complete ignorance. Finally, a CPR inspector with knowledge of the apparatus manages to assemble it. The second team to enter the mine with apparatuses, a group which includes Frank miners G.W. Goodwin, Fred Hallett, William Jolley, Charles Burrows, and Bob Leavitt, enters the no. 1 slant and turns on their oxygen at ten minutes before noon.

With their Fleuss breathing apparatus finally in order, the rescue team from Frank finds three would-be rescuers waiting at the top of the slope, unable to descend because of gas: fire boss

Adlam, who had examined the mine overnight and gone off shift at 7 AM and is now involved in rescue efforts; Abraham Dodd; and another man. The team continues down the slant and moves past five dead men as they go.

Then Leavitt collapses. Either he removes his nose clip, possibly on the inexplicable assumption that he does not need it, or it falls off. Almost immediately, he passes out from carbon monoxide poisoning. The team's inexperience with the breathing apparatus takes over, and some of the team flee back up the slant in a panic. The men who stay carry Leavitt to the top of the slant and put him on the Pulmotor, which Adlam begins to operate. The Pulmotor works for a time, but then must be operated by the men manually, as its oxygen cylinder had run down during the previous resuscitation attempts. Goodwin runs out of the rock tunnel to find a doctor. Men rush in with a stretcher and remove Leavitt from the mine, where doctors are able to revive him.

Outside, reporters ask the rescuers for details of their experiences in the mine. McLeod says he came across men who were found alive, but died before they could be brought to the surface. One miner was found dead sitting with his head in his hands, and many others were found in crawling positions. One survivor was reported to have narrowly missed death because he happened to turn a corner, putting him out of range of the blast. He woke up in hospital.

The apparatus-equipped team in the no. 2 slant returns to the surface, then immediately goes back down again, this time without apparatus. This time Briscoe and the others get all the way to the face, encountering no afterdamp, but a little smoke. The time is after noon. He finds many bodies in the level, and later says it was evident that many of the men had not been killed by the explosion and had moved after the explosion in an effort to find a way out, but had been ultimately overcome by afterdamp.

White, the Bellevue miner, enters the mine at about 1 PM after helping equip the first rescue crew led by fire boss Briscoe. He enters the no. 2 slant without the Fleuss apparatus, and passes, at the no. 1 level, a Pulmotor being used in an attempt to resuscitate victims. Inside no. 2 south level he counts twenty-six dead miners, buckers, drivers, and the timber boss, his cousin Sam Ironmonger. "His watch was going on him; it was 20 past 2 when I got him."[11]

Hours earlier, after waking to the 7 AM mine whistle, George Cruickshank had eaten breakfast and begun his day. As Hillcrest's postmaster as well as a storekeeper, he was sorting and reading his mail when someone knocked at the office door. Cruickshank opened the door to find the Hillcrest Collieries office stenographer, Bob Cherry, standing there in some distress. He told Cruickshank that there had been an explosion at the mine and that Brown wanted him to come up right away. Thirty-eight-year-old Cruickshank knew the faces of Hillcrest men, knowledge that would make him extremely useful at the disaster scene. He rushed up the hill to the mine.[12] Until three the next morning, Cruickshank helps identify bodies whose mine checks are missing. His wife Jean and other Hillcrest women organize coffee, tea, and sandwiches for the rescue crews.

At noon, the train from MacLeod arrives with Constable Wilson, who provides reinforcement for the thinly stretched policemen trying to keep order. They enlist miners to help stretch a wire rope around the mine yard to keep the crowd away from the mine entries and tracks, and ask people to stay behind it. The numbed and passive crowd obeys. Corporals Mead and Grant have taken charge of the wash house, and Constable Hancock the Union Hall. Hancock had come to Canada from England while still a teenager, and joined the RNWMP in 1911. Wilson and the other officers are posted at the entries to the mine to keep the crowds back and maintain order.

At ten minutes past noon, a call is made to the Lethbridge Mine Rescue Station No. 1 to tell them that urgent assistance is required.

Three hours after the explosion, by about 12:30 PM, fans and repaired ventilation have practically cleared the mine of noxious gases, making it easier for rescue teams to search safely for survivors. The exception is the accumulation of more methane inside no. 2 south. The firedamp remains a problem that could lead to more explosions.

Shortly after noon, a CPR special leaves Fernie, B.C. for Hillcrest. The train contains the B.C. mine rescue car and the B.C. government's Draeger rescue apparatuses from the Fernie station. Along the way it picks up all of the available apparatuses from the Crow's Nest Pass Coal Company at Coal Creek, twenty-two trained men acting under the instructions of O'Brien of Fernie Station, Provincial Mine Inspectors Evans and Williams, Dr. Anderson and Nurses White and Dunn of Fernie hospital. It picks up rescuers from Hosmer Mines Limited, and at Michel, W.R. Wilson, general manager of the Crow's Nest Pass Coal Company, with seven trained men, as well as Dr. Wilson and Nurse Brady of the Michel hospital staff.

Among the rescuers on the train from Fernie is a miner from Hosmer named George Loxton. He had worked the midnight shift at Hosmer, showered and changed his clothes, and was eating breakfast when the mine manager told him about the accident at Hillcrest.[13] Loxton was asked to gather other single men who had been trained for rescue, and get ready to leave as quickly as possible. Because of the danger of the rescue work, no married men were allowed to go. When the train arrives at Hosmer from Fernie at nineteen minutes past noon, the rescue team quickly loads the gear onto the CPR special. The train hurtles on through the Pass, through the forests of pine, spruce, and cottonwood, past Crowsnest Mountain, each man lost in his own thoughts as he wonders what lies ahead in the bowels of the mine. They arrive at the Hillcrest station at 2:01 PM, in record time. The officials who meet the rescuers explain the situation as the men load their gear into horse-drawn wagons that haul them up the hill to the mine, where Loxton and the others witness a horrific scene. The women, he later says, are almost crazy with worry and grief, while the RNWMP do what they can to keep order. Loxton's team assembles and dons its two-hour Draeger breathing apparatuses, then enters the exploded mine. They first work on restoring the ventilation.

By 2:30 PM, the Fernie team is fully dressed, equipped and ready to enter the no. 2 slant, but the natural chaos of the disaster takes over: they are told that they are needed instead at the no. 1 slant. They remove their gear and move the heavy apparatus to the other entry, which takes them an extra hour and a half. In the confusion, no one is there to guide or direct them, but they manage to enter the mine at about 4 PM. The team finds bodies, but focuses on repairing the disrupted airways and rebuilding an overcast that had been blown down. As they attend to this, a fire boss appears and orders them out of the mine due to fires in the other slant. They exit, remove their gear, and again pack up their apparatus onto a timber car to proceed back to no. 2, where they are asked to help extinguish the fire. After putting out several small fires, they finally stagger from the mine, exhausted, at 1:30 on Saturday morning. Other members of the B.C. team assist with recovery of bodies, and help wash them in the wash house.

The special train from Lethbridge leaves just before 1 AM with forty trained rescue men from Lethbridge and Coalhurst, along with physicians, six nurses, members of the St. John's Ambulance Corps, and officials. They are all too late. Those who did not escape on their own or with assistance are dead.

CHAPTER 8
CRUICKSHANK'S SHROUDS

And, of course, the dead await attention. While some crews, motivated by hope, search for those still alive in the mine, others repair the airways, and still others bring out the dead. By 2:30 in the afternoon, ten bodies have been brought up. George Frolick, now recovered from his ordeal in the mine, is asked to prepare the wash house for bodies. He in turn asks the carpenters to construct long tables on which bodies will be placed for examination. Miners haul bodies covered with blankets from the mine on lumber trucks and transfer them to coal cars that get pushed to the wash house. There, Corporals Mead and Grant search each body for the brass mine check that will identify him. If the mine check is missing, George Cruickshank and the other miners present try to identify the body. The policeman calls out the check number to the timekeeper, Robert Hood, then removes personal effects such as money, watches, and rings. Hood makes a note of each body's identity and belongings, which are locked up for safekeeping. The policemen, along with volunteers and an undertaker, strip the bodies, wash them, and wrap them in white cotton cheesecloth brought up from Cruickshank's store in Hillcrest.

"The bodies that were brought out were so thickly covered with black coal dust that they looked as if they had been given a thick coat of shoe blacking," explained Cruickshank. "In addition some of them were horribly burnt, and some terribly mangled or dismembered. The clothes first had to be cut away with shears, and then the bodies thoroughly washed. So quickly did the work have to be done to make room for the bodies that were being brought out next, that there was no time for fastidiousness. They were sluiced off with sprinklers and then rubbed with rough towels."[1] Men both living and dead crowd the wash house as the hours pass: doctors, RNWMP officers, volunteers, and company officials. Among those who watch is Jonathan Penn, whose father and brother were in the mine at the time of the blast. He looks for their faces among the dead, and is startled when he observes that some of the victims appear to shiver as though alive. A doctor explains that air inside the lungs of the dead men causes the tremors, and that despite appearances, they are indeed deceased.

The work in the wash house is made particularly unpleasant by the state of many of the bodies, which have been torn to pieces by the tremendous force of the explosion. Limbs, heads, and fragments of torsos have to be set aside and pieced together. Many of the volunteers numb themselves with alcohol.

Union members watch the work of the RNWMP officers closely and suspiciously, but they gradually replace their deep-seated mistrust of the police with a grudging admiration. A few days later, Inspector Junget, in a letter to Superintendent Starnes at MacLeod, will make a plea for special recognition for Mead, Grant, and Hancock. He comments that their work received praise not only from citizens, but also from officers of the miners' union, "some of them being the rankest socialists and the enemies of any Police or Military Forces." In the letter he quotes union offi-cials as saying the miners "have no use for the Police but cannot help respecting its members when we see them work under such trying conditions." Later, each of the three men is granted $50 by the federal government in recognition of their outstanding ser-vice at the Hillcrest disaster.

Among the rotating teams of miners who help wash and wrap the dead are survivors Yeadon and Frolick. They work early into the next morning at their gruesome task. From the wash house, volunteers carry the bodies outside, wrapped in shrouds, and lay them into freight wagons owned by Jack Eddy, a logging contractor, rancher, farmer, and owner of Hillcrest's first livery barn. He and Rinaldo "Joe" Fumigalli transport the bodies down the rough road from the mine site to the hospital, the union hall, the Masonic hall above Cruickshank's store, and the boarding house, where they are laid out on the floor in rows.

Miner George Loxton's rescue team from Hosmer first helps restore the ventilation, then recovers bodies, but before he and his team are allowed into the mine they must write their own wills. Loxton wills his possessions to his brother, who is also among the rescuers. Just before they enter the mine, other miners offer them a shot of brandy to bolster their courage. With their breathing apparatus on, Loxton and his teammates penetrate deep into the mine, where they are faced with the dead. Recalling the experi-ence afterward, Loxton said: "Up in the places there, there was no damage done, but when we got up in there, we found men. They're sittin' there, dead. Some of them never knew what had happened, I suppose. They had heard the explosion, and then the gas had killed them. And the trouble was gettin' those men out because their bodies were rigid, you see." What Loxton implies is that in order to remove the bodies, they often had to break the bones of bodies fixed into awkward poses by rigor mortis. What disturbs Loxton the most is coming upon dead men that he knew: "And going up in there, you know, you'd go into a place, and see a man sittin' down or something, and you knew him. I knew some of these men. And you know that kinda makes you feel funny. Of course, you gotta put up with that."

Loxton and his coworkers are in the mine for two hours at a time, which is as long as their breathing apparatus will allow. But the weather has turned cold, and when they exit the mine, they are forced to retreat for warmth to the wash house, which was the only building with heat. "And you know, those bodies was burned, and the smell of that was terrible, and we had to stay there because it was warm. If we went out... perspiration, everything... we'd freeze."[2] The only other place to get warm is a bonfire built out in the open in the mine yard. There men sit on a nearby pile of rocks to warm themselves, some wearing nothing but boots, pants, and thin undershirts.

Women in nearby communities, most from mining families, have rushed to assist the rescue efforts and relieve families affected by the disaster. When the local food supply is exhausted, loads of bread, butter, milk, and the like are rushed in from other camps. Residents of Blairmore send in six hundred loaves of bread, while another six hundred arrive from Coleman. Women from Bellevue and other towns are on the scene within the hour with refreshments for the rescuers: galvanized buckets, even washtubs, full of coffee and tea.

By the time Stirling arrives at Hillcrest on the train from Lethbridge, between three and four in the afternoon, hope has generally been abandoned that anyone is left alive in the mine. All that remains to do is recover the bodies. About twenty bodies have been recovered so far. Stirling makes a quick assessment of the situation, then wires a telegram to Stocks. He estimates the number of dead at 196. The cause, he says, is unknown. His job, until the bodies are recovered, will be to update Stocks on the situation with as much detail as he can, including estimates of the number of dead. During the days ahead he will wander the mine and try to figure out what caused the explosion. He will take copious notes, and during his inspections with various fire bosses will come to know the mine intimately.

With the Lethbridge mine rescue team now on site, officials send the team of Ed Berford, Aleck Stevenson, A. Quinn, G. Coutts, W. Goldie, G. Hargreaves, and Sam Jones, led by John Stevenson, into the mine immediately. That same afternoon, a special train arrives from Calgary with CPR officials, including the General Superintendent of the Alberta Division, the general manager of the Natural Resources Department P.L. Naismith, a publicity agent, a detective inspector, CPR constables, a party of newspapermen, and two physicians picked up on the way. The party is taken from the Hillcrest station to the tipple by the Hillcrest locomotive, Old Maude. The doctors on site now number about thirty, with perhaps fifteen nurses. However, there is nothing for the medical personnel to do except determine the causes of death. The victims of carbon monoxide poisoning are easily discerned once their bodies have been cleansed of coal dust. These bodies are characterized by a pink flush, which gives them a lifelike appearance.

One of the miners on a recovery team, J.R. "Doc" McLeod, enters the mine determined to find his friend, twenty-four-year-old miner John Hood. Doc enters the mine, but is able to descend only a short distance when his head begins to swim and he becomes dizzy and weak. He leans against the wall and begins to vomit. Those with him rush him to the top, where he recovers. Again he enters the mine, but this time does not stop until he gets about 300 feet into the mine. He finds Hood dead on the floor, lifts him up, and carries him back to the surface and into the sunshine. Hood's body is slightly burned on his upper body and face, and has suffered lacerations, but he survived the blast — he died of carbon monoxide poisoning while trying to escape.

Fire boss Adlam re-enters no. 1 at about 3 PM with a breathing apparatus, and stays in for half an hour. He and two others find a body. They go as far as the old south level. He returns to the top, but then descends again and works to restore the stoppings until early the next morning. No rescue parties enter no. 1 while Adlam is working.

At 4 PM, top pump no. 2 is restarted, and begins pumping water from the mine. At about the same time, Bellevue miner White finds fires. The loose coal on the pavement is burning. White packs water along no. 1 south from a hole filled with water on the level, then hunts for his brother along no. 1 south.

By 8 PM, fifty-two bodies have been recovered, and work continues into the night.

District Inspector Francis Aspinall arrives in Hillcrest and enters the workings before midnight.

At 11:30 PM the rescue teams all are recalled after one party encounters another fire. Officials send the rescue party from Hosmer, led by Shaw, down to investigate before any more rescue parties are allowed down. They do encounter fire, and immediately send a call to the surface for more men and water. Only the already-exhausted Lethbridge crew comes forward. The fires are small, but prove difficult to extinguish. Stirling counts fourteen small fires in one place. "The procedure adopted at first being to dig them out and latterly to extinguish them by means of water carried from the discharge of the pump by means of fire hose," he later explains.

As day breaks on the morning of the second day, Saturday, the Coroner's Inquest meets with a jury of ten men sworn in by Coroner Pinkney. Pinkney is the same coroner who had headed the inquest into the Bellevue disaster in 1910 when he lived in Lille. It was a long, exhaustive inquiry, and led to major changes in Alberta's mine rescue preparedness.

Originally from Darlington County, Durham, England, Pinkney had arrived in Frank in July of 1905 to take up a job as accountant with the company building a lead ore/zinc smelter at Frank. Shortly after his arrival, Pinkney became involved in plant operation, and lost his right hand in an accident at the smelter. Until his accident, he had also been an accomplished musician. With the loss of his hand and his job, Pinkney's life took another turn. In 1908 he was appointed coroner for the area from Pincher Creek to Coleman. Later, from his office in Blairmore, he also offered his services as a commissioner for affidavits, an insurance and commission agent, and an agent for the Winnipeg Piano Company. The latter's slogan was "Organs, Pianos, and Talking Machines sold on easy terms, and a liberal discount for Cash."[3]

Now Pinkney's job is to examine, along with his jury, the bodies of the Hillcrest victims and determine the cause of death. The jury consists of a blacksmith from Blairmore, James W. Gresham, who is chosen chairman and was also the chairman of the coroner's jury in the Bellevue disaster; a hotelman from Hillcrest, Charles Fuchs, who is the proprietor of the Union Hotel; a gentleman from Frank, Arthur E. Fariner; and miners John J. Thomas of Frank, Harry Smith of Coleman, Thomas H. Duncan of Passburg, and George Grafton, Evan Ross Mackenzie, G. William Goodwin, and John Shorre of Bellevue. Goodwin had previously been on the first rescue team to enter slant no. 1 with rescue apparatus.

Pinkney and his jury view the bodies recovered so far, and conclude that they were killed as a result of an explosion of gas and coal dust. Eventually, they will hold a hearing, call witnesses, and take testimony. The formal session of the inquest is adjourned until July 2, but in the interim the jury inspects bodies as they are recovered. By 10 AM a total of eighty-six dead bodies have been taken from the mine and brought to the wash house from the workings in mine cars, two at a time.

Stirling has been awake at the mine all night. In the morning, he sends another telegraph to Deputy Minister Stocks. At 6 AM, Stirling meets with the leaders of the B.C. rescue contingent and others in charge of the recovery operations. He tells the B.C. team that the situation is well in hand, and that they can all go home. They catch the morning passenger train, but at Hosmer hear a rumour that another explosion has occurred, and that a rescue party is trapped. As tired as they are, the men volunteer to return to Hillcrest. They wait, ready to depart until word comes via telegraph at Fernie that there has been no second explosion, no one is trapped, and they are not needed after all. Only then do they resume their journey back to Fernie.

In Hosmer, the miners involved in the rescue, after being awake for forty-eight hours, are given enough liquor to get them drunk. Loxton then sinks into bed and sleeps for twelve hours.

RNWMP Superintendent Starnes, unaccustomed to watching action from the sidelines, and finally relieved at MacLeod, boards the first morning train to Hillcrest. Inspector Junget waits for him on the station platform. The two men shake hands, and Junget suggests that the Union Hotel's bar should be closed, as the amount of alcohol being consumed by local miners was a problem that was only getting worse. The small bar fills up quickly, and soon other groups of men, unable to get in, gather outside. Those inside pass glasses of whiskey to the men outside. Drunk miners are no help whatsoever in rescue or recovery operations, to their families or to their employers. The battered town doesn't need this new problem. This was all Starnes needs to hear. He steps into the station's small CPR telegraph office, and an operator taps out a message to the deputy attorney general in Edmonton. The response is immediate and affirmative. The liquor license at the Union is suspended until Monday night.

The two policemen climb into their automobile and drive while Junget reports on events and points to the mine tipple on the ridge above the town. They drive across the Crowsnest River, up the rough road past the cemetery, where men and horses have begun to scrape out two huge graves in the rocky earth, past the Cruickshank-Burnett store on the left and the miner's hall, where bodies are laid out and coffins are stacked, towards the Union Hotel. As the Mounties approach the clapboard hotel, the crowd of boisterous men gathered in front fall silent. The Mounties nod politely at them, then disappear into the shadows of the busy hotel. Inside, they summon the surprised owner, Charles Fuchs, a member of the coroner's jury, and tell him that his license to serve liquor has been suspended until Monday night. Fuchs immediately steps into his bar, the policemen at his elbow, and announces to the assembled miners that he has to close the bar on orders from Edmonton. He locks the door behind the last man out. The miners disperse, many to the hotel in Bellevue, where they continue their libations. Starnes then heads up to the mine, where he meets with several of the provincial mines officials and CPR officials on site.

At the mine, recovery operations continue, while a numbed calm descends on the crowd. Hundreds of people line the hill-sides along the tracks, and anywhere else they can watch the activity at the mine's two exits. The men gathered at the door-way of the wash house talk quietly among themselves. Two men with impassive faces push a mine car with a huddled, indistinct form inside near the wash house. All eyes watch without emotion as an RNWMP officer and several miners struggle to lift the twisted, blanket-shrouded shape from the car and carry it down the wooden steps, through the crowd of men at the door, and into the wash house. For a few minutes, only the sound of the hoist can be heard as it pulls another trip to the surface. Then the men return for the next mine car and its heartbreaking cargo.

At around 9 AM, Wilde enters no. 1 with rescuers from Pass-burg and Michel, including the superintendent at the Passburg mine, Joe Thomas, who is in charge of the party, and stay in the mine until about 2 PM. The breathing apparatuses are no longer needed. Later, Fire Boss Adlam also re-enters the no. 1, and meets up with the Thomas party. They find the bodies of Jack Clarke, Bob Stratton, Galley Moore, and John Davidson, among others.

By 6 PM on Sunday, 177 bodies are recovered, with twelve still inside the mine. By midnight, eight more bodies are recovered, leaving only four unaccounted for. None is recovered through Monday and Tuesday.

Late on Tuesday, June 23, Stirling, who had spent a good deal of his time underground at Hillcrest, telegraphs Deputy Minister Stocks in Edmonton to report the most recent numbers. All of the bodies are buried, even though nine were never identified. Stirling adds that he is fairly satisfied that the explosion was not caused by shots firing.

That afternoon, the fires are finally all extinguished.

The next day, Hillcrest Collieries President Mackie arrives on the train from Montreal and immediately goes into conference with mine officials. After the meeting, manager Brown and other officials begin a thorough examination of the mine that lasts through the next day, while recovery and mine repair efforts continue.

On Friday, the company releases a list of the names of 186 recovered bodies. The three remaining bodies are said to be Oakley, Bainbridge, and an unknown.

CHAPTER 9
THE REPORTERS

At the Hillcrest Railway station, reporters swamped the four telegraph operators with press dispatches about the disaster to their newspapers and inquiries and questions from afar about the extent of the disaster, as well as the safety of particular individuals.

Among those who rushed to the mine were newspaper reporters and photographers from towns in the Pass, Lethbridge, Calgary, and Edmonton. The reports in the *Fernie Free Press, Coleman Bulletin, Edmonton Bulletin, Lethbridge Daily Herald, Medicine Hat Daily News, Bellevue Times, Blairmore Enterprise, Calgary News-Telegram, Morning Albertan,* and others are characterized by gross exaggerations, fabrications, and a general lack of facts. One inventive reporter quoted an unnamed survivor as saying, "It came upon us like a huge breath of coal dust, flying chunks and gas. There was no report that I heard. Just a dull rumble and a horrible, black, blinding, choking hurricane, the intensity of which seemed to grow every second like a huge growling monster of destruction."

The supposed quote offers evidence that at least one reporter on the ground at Hillcrest, A.C. Yokome of Calgary's *Morning Albertan,* was not above some hack wordsmithing to entertain his readers. The present-day reader isn't sure whether to laugh, question the truth of all other quotes, or both. Of course, no reporter could get everything wrong. The difficulty lies in separating fact from fiction — a subjective process at best.

Both the *Lethbridge Daily Herald* and *The Medicine Hat News,* for example, reported that the engineer at the no. 1 hoist house had narrowly escaped death "when the building was completely demolished." This demolition was greatly exaggerated. The wall facing the no. 1 entry was blown down and the roof demolished, but when the mine closed in 1939, the same building was still in use. As if that exaggeration weren't enough, the report continues, "It is stated that a small boy, who was in the building at the time, was blown to atoms." Blown to atoms? If he were blown to atoms, then not even pieces of him remained, and it would be difficult to prove he was ever there. Apparently the newspaper assumed that readers would not ask if it was credible that a small boy was inside the hoist house. It made for good copy, with a tone of emotional horror to it, but was completely false. No child died, or was even injured, in the Hillcrest disaster.

Of course, not all newspapers and reporters were responsible for these inventions. The *Calgary Daily Herald* stood out for the accuracy of its reporting, as did the *Coleman Bulletin.* Even the *Herald,* however, in its first report of the disaster grossly misrepresented the numbers of men involved: its subhead declared that 250 men had perished.

Certainly, reporters would not have had an easy job gathering facts. In one photograph taken at the rear of the union hall, where relatives and caskets are gathered, and where bodies were laid out on the floor for identification and viewing, two men glare with hostility at the camera and the photographer. Union members, whether from Hillcrest or from other camps, would have been protective of fellow miners and their grieving relatives. Mine officials, on the other hand, were busy with rescue and recovery, and many miners were exhausted from days without sleep. Yokome himself writes, "It was only during breathing spells in the relief work that *The Albertan* was able to get attention of the survivors for a sufficiently long period to draw from them their harrowing experiences."

A report in the *Calgary Herald* from June 20 similarly remarked, "The officials of the company... are so busy looking after the work of bringing out the bodies and other matters or so worn out with labour and want of sleep that they are in no position even to discuss what little is known of the cause of the explosion." The report adds that the little information available was collected in fragments and by direct observation. The honesty of these remarks, written by a reporter described only as a "Herald Staff Correspondent," make this particular report more trustworthy than some others.

Just as men who return from ferocious battles in wartime are seldom willing to tell of their experiences, so too were the rescuers at Hillcrest. A report in the June 22 issue of the *Calgary News-Telegram* stated that "[s]everal of the rescue party from Lethbridge who worked many hours in the mine securing the bodies of the entombed miners were asked to relate their experiences while in the mine, but without exception they asked to be excused, except for a short statement." The statement said only that they had found bodies, some face-down and others stretched out against the wall, and that the fumes were bad. It was a statement of the obvious, in keeping with the mental state of tired men. Their refusal to answer questions, however, would have been a crushing disappointment to the assembled reporters.

To meet demands for copy, some reporters were not above speculation. A small item in the June 23 issue of the *Calgary News-Telegram* presented a new theory as to the cause of the explosion. A man described only as a "Pit Boss, of another mine" was reported to believe, based on alleged reports of three distinct explosions, "that the first was caused by gunpowder used in blasting, the charge of gas exploding by concussion rather than ignition." Had the reporter actually done some investigation and asked mining experts what generally causes mine explosions, he might have helped educate the public about coal mining, but instead only helped fuel the pointless speculation, which in turn helped to fuel the blame.

On June 19, the *Fernie Free Press* reported, "220 Miners killed in Explosion at Hillcrest. Only Thirty Men Escape Alive Out of 250 who Entered the Mine Last Night." The explosion was reported to have taken place at 7 AM, the result of "a sudden outburst of gas, which ignited dust penetrating to every part of the workings." The paper concluded its report with the erroneous claim that the Hillcrest mine was owned by the Canadian Pacific Railway. Apologizing for the lack of information, the newspaper explained, "Owing to the rush of private messages it is impossible to obtain much information up to the time of going to press." The report got the number killed wrong, the number who entered the mine wrong, the time of the explosion wrong, the cause wrong, and, finally, the ownership of the mine wrong. Virtually the only thing the report got right was there was a disaster at Hillcrest.

The front-page headline in the June 19 *Medicine Hat News* screamed "225 DEAD: HILLCREST MINE DISASTER." An additional 300 men, it reported, were entombed inside.

The *Lethbridge Daily Herald*'s headline on June 19 said 225 men were entombed, 275 men were in the mine when it exploded, and twelve bodies had been identified. Among them was the name of Sydney Bainbridge, an impossibility, as his body was never recovered from the mine.

The *Bellevue Times* reported on the day of the explosion that upwards of 200 men had been killed. A week after the explosion, on June 26, the newspaper reported that "First reports of the extent of the disaster were, we regret to say, not at all exaggerated, for it is generally conceded to have been the worst disaster in the world's history as far as the percentage of population affected is concerned." The absurdity, even the arrogance, of the *Bellevue Times*'s claims aside, the *Calgary Herald* accurately reported as early as June 22 the number of men in the mine at the time of the explosion, the number of men rescued, and the number killed — this despite the fact that the official list of victims would not be handed out to reporters until June 24.

The June 27 issue of the *District Ledger* at Fernie reported correctly that 189 had perished, and also commented on the accuracy, or lack thereof, by the daily press who were true to form, the article claimed, with sensational and inauthentic accounts of survivors' experiences. The report added that theories allegedly advanced by survivors could be disregarded, and that no practical mining man would express an opinion until all the facts were presented at a proper inquiry. With some insight, the report pointed out that the effect of the explosion on survivors was to numb every faculty to the point where survivors could not explain how they reached safety, could not remember the conditions of the travelling road, and expressed stories completely at odds with the men next to them. The report adds that such variance in accounts can only be understood by men who have had previous experience with such disasters.

In a similar vein, a reporter for Calgary's *Morning Albertan* recounted on June 20 that he had "interviewed no fewer than 15 survivors, but not a single one of them had a comprehensive idea of what had taken place." The report continues, "In the horror of the terrible calamity they temporarily lost their powers of observation and reasoning and remember nothing but the loud report and the hail of pelting rocks, coal fragments, and storm of coal dust."

Certainly the most egregious fabrication concerns the death of David Murray, Sr., who reportedly escaped from the mine after the explosion, then fought off a Mountie to run back in search of his three sons, and died in the mine along with them. The truth of this tale, which has become one of the most frequently told stories of the disaster and its aftermath, deserves serious scrutiny based on the evidence, or lack thereof. The first mention of it occurs in a report in the *Lethbridge Herald* on June 20, 1914: "Chas. Murray gave up his life in an attempt to save his two boys, who were entombed in the mine. Hearing of this, he rushed in but never came out again." Not only is Murray's first name printed incorrectly, so is the number of his sons in the mine. The *Herald* repeated the story on June 27 with slightly more detail, and alleged that the remains of all four men were taken to Fernie for burial. In fact, they were buried at Hillcrest. The story of Murray's alleged rush back into the mine was also reported on June 25 in the *Coleman Bulletin*. However, a significant number of newspapers, including Alberta's largest circulation papers, Calgary's *Morning Albertan,* the *Calgary Herald,* and the *Edmonton Bulletin,* did not report the alleged event. The *Albertan* even reported on the deaths of the Murrays and the terrible effect upon the widow, but made no mention of the manner of David Murray, Sr.'s death as alleged in the *Lethbridge Herald*. Nor was the incident reported by *The District Ledger,* which in fact commented with a note of irritation on the specious reports and claims being made by other newspaper reporters. Perhaps more telling is the fact that the police failed to mention it in any of the RNWMP reports made by officers on the scene to their superiors, and that it was never spoken of during the inquiry or the inquest.

Frank Anderson embellished the story considerably in his account of the disaster, first published his 1983 book *Tragedies of the Crowsnest Pass.*[1] In this account, Constable Hancock is portrayed as the RNWMP officer who wrestled with Murray in an attempt to keep him from re-entering the mine. Hancock died in 1975, so Anderson could not have spoken with him about his role in the disaster unless he did it at least eight years before his account was published. To add to the doubt, Hancock was a fairly large man; David Murray was not.

The facts of the matter, as found in company and Alberta government records, speak volumes. David Murray, Sr. died of carbon monoxide poisoning, as did his mining partner Dan Cullinen. Their bodies were found near the bottom of the no. 2 slant at the third level south, right at the face where they had been working. It seems extremely unlikely that Murray got out from the bottom of the no. 2 slant when other miners far higher up had barely made it out alive, and then ran back in, all the way back down the no. 2 slant, to die exactly where he had been working.

Unfortunately for his widow, David Murray was simply a victim of circumstances, just like all the other men who died in the Hillcrest mine. These facts are at odds with the newspaper reports, which almost certainly invented the manner of Murray's death, or else repeated it without checking the facts.

On the other hand, when reporters described what they observed, not what they fabricated, the pictures they drew could be chilling. The June 20 report on the front page of the *Edmonton Bulletin* illustrates the power a reporter could summon when he ignored the urge to slap pancake makeup on the incident. "Some of the bodies are bruised, torn and broken as though the force of the explosion had driven them many feet in the air. The copy details the discovery of bodies by the rescue gangs in the places where the explosion was most severe: "here were picked up bodies burned to a crisp — torn to pieces and picked up in fragments, decapitated bodies, others with their clothes completely torn from them, and others so stiffly rigid that their arms or legs had to be broken in order to allow their corpses to be enclosed in the caskets." Here is gruesome, unembellished fact that paints a word picture of the immense power of the explosions that both miners and the public needed to understand.

Sometimes the reporters, despite their best efforts, managed to evoke images that linger in the imagination. The June 22 *Calgary News-Telegram* headed a column: "FEW BODIES NOW TO BE RECOVERED. Searchers Have Brought 181 of 189 Dead to the

Surface." The report describes the start of the burials, and comments on the stacks of coffins. "On Sunday night several carloads [boxcars] of caskets were unloaded at the Hillcrest depot and on Monday morning other consignments arrived and were piled in the centre of the mining town, until it looked like a huge store yard for undertakers' supplies."

The newspapers' often mawkishly sentimental writing style is certainly not in keeping with present-day tastes. In the June 22 *Edmonton Bulletin,* for instance, a writer describes the activity outside the union hall, where bodies lie in caskets: "Widows were led away from the last fond gaze of all that is mortal of their husband, and the moist eyes of onlookers were not a few. It was not infrequent that the lids of the caskets were opened and kisses imprinted on the cold lips of loved ones. One full blooded young man busily engaged in the work of encasing the dead, quickly lifted the lid of a grey casket containing the remains of his father, pressed several kisses on the silent lips, then murmuring, 'Good Bye, no more.'"

A reporter for the *Coleman Bulletin* also sought an emotional response when, on June 25, he told of an experience on a street in Hillcrest that was unusually crowded with strangers: rescue workers, church ministers, miners from nearby communities, reporters, undertakers, aid workers, CPR officials, government, company and union officials, and the simply curious. The reporter tells of seeing a young girl on the path who wept "in a most heart-rending manner, the while gazing upon a ring on her finger, her action plainly telling the story of a sweetheart snatched from her by the hand of death."

In a similar approach, in the same story, which we must assume is factual, the reporter asks us to look upon the innocence of childhood from the position of adults who know the meaning of pain that the child cannot yet grasp. He describes an encounter with a little boy of about seven on the street.

The child approached the reporter and asked him, "Do you think my daddy will be home by morning?"

"I hope so, my little man," was the only answer the reporter could think of, and then advised him to go home and wait until morning.

"No, I'm going up to the mine to find mama," the child said. The helplessness of the reporter in confronting innocence in the face of horror, the knowledge that the child's father would never come home, and the desire to let the child hang on to hope tell a powerful story in just a few lines.

The reporter continues, describing newly married fire boss Sam Charlton and two other young brides "of little more than a week who have been so suddenly widowed."

Often, he wrote, "a stretcher would be carried out from the mine entrance bearing a body mutilated beyond recognition. Sometimes there was only one leg showing. Occasionally the body was decapitated, and sometimes the stretcher bearers carried a burden too horrible and gruesome to describe."

The reporter comments obliquely on the power of routine and expectations to comfort us, and which, when broken, jolt our security. He describes the behaviour of the women and others who waited on the hillsides beside the entries. From the time the explosion occurred until about three PM, the time at which the day shift always left the mine, the women and men who waited displayed marvelous self-control. But after three o'clock, the catastrophe struck with added force. Grave but now silent groups of men stood scattered over the mountain, while women wept, clung to each other and tried to comfort their children, who had suddenly become frightened.

The *Morning Albertan* on June 20 carried news of the disaster on the front page. Yokome reported that "[m]any of the bereaved are loath to leave their homes. Overcome with shock and grief, they sit on the verandahs of their humble cottages swaying to and fro in silent pain or moaning and sobbing in hysterical grief. For the most part the mourners are subdued in their expressions of sorrow, their agony too deep for superficial manifestation."

That reporters found anyone to question aside from the town fool is a minor miracle in itself. Chaos reigned, and out of the confusion they had to find information to pass on to their demanding editors. And so what was often reported was the gist of events, if not the hard facts.

The only other source of news for the general public was motion picture theatres. But because of the slow speed of distribution, the lack of film processing facilities, and the scarcity of movie cameras and film stock, the footage of supposedly newsworthy events would not be screened until a considerable time after the events took place, by which time the newspapers had moved on to other, newer events. The public, however, remained fascinated by the movies, and gladly watched these non-fiction news films anyway.

From an historical point of view, one of the most disappointing aspects of the post-disaster events at Hillcrest was the presence of a photographer who owned a movie camera but refused to shoot footage of the events. The talented Lethbridge photographer Arthur Rafton-Canning had rushed to Hillcrest with his movie camera upon hearing of the disaster. Rafton-Canning had immigrated to Canada from England in 1880 and had served as a private in the Queen's Own Rifles, 2nd Battalion with the North West Field Force during the Riel Rebellion. In 1907 in Lethbridge, he established the British & Colonial Photographic Company. Very little of life in southern Alberta escaped his lens, and the cherished record of his photographs now rests in various Canadian archives, but he could not bring himself to photograph the events at Hillcrest.[2] Shortly after this, he left Alberta. Rafton-Canning's observation of the RNWMP in action must have excited him, however, for that same year he argued that the RNWMP should make use of moving pictures in the future.

Another young man with a camera rushed to the mine, but unlike Rafton-Canning, Thomas Gushel did not hesitate to photograph what he saw. It was the beginning of Gushel's long career as a professional photographer in the Crowsnest Pass. Today, most of the few photographs we have of the disaster's aftermath are attributed to Gushel, and his photographs of the Pass, the miners, their families, and other Pass residents depict a rich part of our history.

The quickest way to present news in theatres was with the projection not of movies, but of still photographs in the form of magic lantern slides, which were black-and-white or hand-coloured glass slides. At least two theatres, one in Calgary and one in Lethbridge, advertised screenings of photographs taken at the disaster scene. The Empress in Calgary advertised on June 23, "IMPORTANT SCENES TAKEN FROM THAT HORRIBLE HILLCREST MINE DISASTER." The ad in the Calgary Daily Herald continued, "Only 48 Men Saved Out of a Total of 247; GENUINE PICTURES OF THE WORLD'S WORST MINING CATASTROPHE, SHOWING FOR TWO DAYS ONLY AT USUAL PRICES OF ADMISSION." The Starland in Lethbridge, meanwhile, advertised in The Lethbridge Daily Herald, "TO-NIGHT, Scenes Taken from that Terrible HILLCREST MINE DISASTER. These Remarkable Pictures will Be Shown Tonight in Addition to An All Feature Bill." These slides have since vanished, but if they ever reappear, they will tell a story that is largely missing from the chronicle of the disaster.

The Calgary Daily Herald used its pages to solicit donations for its relief fund. One of the solicitations was headed, "Hillcrest Miners' Families No Race Suicide Aggregation — Relief Is Greatly Needed." Some members of the upper class feared that Anglo-Saxons risked committing "race suicide" by using birth control and failing to keep their birth rate up compared with that of immigrants and minorities. The idea and the term began in 1891 and continued through the 1920s. The Herald's request for donations commented, perhaps tongue-in-cheek, "There is no such thing as race suicide in the homes of the miners at Hillcrest. Following the disaster of last Friday many families of six, eight and ten children were left without a father or breadwinner of any sort other than mother, whose ability in that role is meagre, and whose opportunity is practically non-existent."

The piece also reminded readers that the disaster was an Alberta affair, and Albertans ought to look after their own. The writer urged Albertans to reject outside aid, should it be offered, as rumours suggested it might be from Britain. This comment reflected the fierce pride and independence of Albertans.

A second call for donations appeared in the next day's paper, this one headed, "The Spirits of Dead White Men Call to the Living — Are You Listening?" The plea continued, "It's a white man's job, for the Hillcrest camp was a white man's camp. Put yourself in the place of one of these white men, cut off without a moment's warning and with wife and children dependent upon you for support."

The Hillcrest mine had been described in a number of newspaper reports as "a white man's camp,"[3] which reflected the fact that only Europeans worked in the mine. Chinese, in particular, were excluded in keeping with the demands of the UMWA. Perhaps these workers could take consolation from the fact that their exclusion from the mines excluded them from mine tragedies as well.

CHAPTER 10
THE DEAD AND THE BROKEN

Dead bodies seemed to lie everywhere: at the hospital, at the union hall, and in the Masonic lodge above the Cruickshank store. And the urgency of their disposition took over other activities: farewells would have to be swift. R.E. Campbell and George Cruickshank supervised the identification and classification of the dead miners. On Saturday morning, two railcar loads of coffins arrived from Calgary and MacLeod, and at the Hillcrest cemetery a crew of some forty men with equipment supplied by the CPR had been put to work digging graves in the rocky earth. The next morning, at the miners' hall, the undertakers, assisted by miners, placed the bodies in coffins and carried them down and out to the nearby common area next to the Cruickshank store. There, Cruickshank and Campbell numbered each coffin with chalk, and beside each number wrote the name of the dead miner inside. In an age when faith played an important role in most men's lives, they placed the caskets side by side in rows according to their faith, opposite letters that stood for the various denominations:

"A" for Anglican, "C" for Roman Catholic, and so on. A throng of about 600 mourners gathered on the green where the coffins lie in a roped-off area. Distraught family members wandered among the caskets in a search for the remains of their men. Some mourners unscrewed the lids of the coffins to see loved faces for the last time. Someone informed one of the mourners, a miner who escaped the disaster, that according to reports he had been killed, and therefore should not be walking around. Pat Cossimo was at first shocked, then felt a grim sense of amusement when he was shown a casket with his own named chalked on it.

At 11 AM the Masons gathered, two hundred strong from Mac-Leod, Pincher Creek, Cowley, and Regina, joined by the eighty members of the Sentinel Lodge for the first of the funerals. The weather failed to cooperate, and as if in concert with the gloom of the mood a cold, harsh wind began to blow. As it raked the ground it swept up fine coal dust and hurled it into the stinging, already reddened eyes of mourners. The clergymen held simultaneous services at 1 PM in the Anglican church for the Anglicans, and in the Methodist church for the Methodists, Presbyterians, and Baptists.

Later, under a thick, leaden sky, men lowered the coffins into their rough wooden boxes, and loaded them into horse-drawn wagons that had been gathered from nearby towns. At 2:30 PM, the wagons began the slow procession to the graveyard, led by the Bellevue and Michel bands who played the "Dead March" from Handel's opera *Saul*. Mourners followed the wagons on foot to the nearby cemetery. There, the Bellevue and Coleman miners' bands, as well as the Crows Nest Pass Italian band, played as the wagons stopped, the men removed the shells and placed them on the ground, and the wagons moved on to make room for the next trip. Other men lifted the caskets and slowly carried them down into two large trenches, each about fifty yards in length, and placed them side by side. The Catholic victims were laid in lines in one trench, with spaces left for others yet to be buried. In the other, mourners laid members of the Anglican, Methodist, and

Baptist denominations. Members of the lodges — Masons, Odd Fellows and Orangemen — were laid in their separate plots. Some caskets were carried to separate graves that had been dug at the relatives' request. Thomas Corkill's relatives buried him at Fernie.

On this Sunday, the mourners buried 117 men in the Hillcrest cemetery. The last mourner stayed at the graves until after dark.

Meanwhile, teams of miners worked urgently to extinguish the fires in the chill tunnels, and by Sunday afternoon, they believed that they had finally succeeded, and recovery teams could resume their work in relative safety. The next day, however, Stirling wired Stocks to tell him that he was arranging to have water pipes laid through the mine to cope with the fires that were still, in fact, causing trouble.

With the situation now in hand, Superintendent Starnes returned to his post in Fort MacLeod. That evening he received a telephone call from Inspector Junget, who told him that 177 bodies had been recovered by 6 PM, and that twelve others were supposed to be still in the mine.

For Isabella Petrie, the disaster could not have been more devastating. Upon her husband James's death in Scotland in 1902, she had been left with eight children, aged six to sixteen, to raise on her own.[1] Together they crossed the Atlantic in 1906 to join her sister and her sister's husband in Blairmore. There, in a desperate attempt at financial stability, she opened a boarding house in a two-storey log building. It was a success, and the family moved to a larger building where they could take in more boarders. Isabella's hard work paid off, and eventually, in 1912, the family moved to Hillcrest, where she took over a store and small restaurant, and once again took in boarders. When her sister passed away, she married her brother-in-law. A daughter from this marriage died in 1914, but in fact her troubles had just begun.

On the morning of June 19, her twenty-two-year-old son Joseph came off shift at 7 AM, but three other sons — James, 28, Robert, 24, and Alex, 17 — had entered the mine for the day's work. In the wake of the explosion, Isabella had watched anxiously as the bodies arrived in wagons, and were then carried upstairs to the miners' hall above her store. The bodies of Isabella's sons were among the dead recovered. Another son, Andrew, had been scheduled to work the next day. Prostrate with grief, she collapsed and was rushed to hospital.

Fire Boss Sam Charleton had been married just a month before the disaster, and at the same time had been installed as a lay reader of Hillcrest's Anglican Church.

Miner David Murray, a Scottish immigrant, and his three sons, David, William, and Robert, were all killed in the mine. Before they moved to Hillcrest, Murray, his wife Elizabeth, and their ten children had lived at Coal Creek, near Fernie. On August 1, 1908, the family had been on a train on their way back to Scotland when they found themselves caught in the great Fernie fire. The inferno burned the town of 6,000 to the ground, killed ten people, destroyed 1,000 buildings, and left only twenty-three houses standing. The Murray family had fled the train, which had caught fire, and barely escaped with their lives. They continued back to Scotland, but soon returned to Canada, finally convinced that however bad conditions were in Canada, they were worse in Scotland.

Some of the men killed were not even miners by trade. Fred Bennett and William Fines, rather than remain idle, had accepted jobs in the mine. Bennett had been a soldier with service in Egypt and India, and had visited many parts of the world. Fines had been in Canada about a year. After his arrival from Scotland, he had worked for the most part as a plasterer. Frank Bostock of Illchester, Nottingham, England, was a soldier, not a miner. He had served with distinction in the last Boxer war in China, Malta, and Bermuda, but died far from combat in the Hillcrest mine. Soon they would lie among miners in the Hillcrest cemetery.

The people of Coleman keenly felt the savage bite of the disaster. Among the dead was Paddy Kane, a Coleman resident of about nine years who had begun work at Hillcrest some fifteen months previous. The forty-one-year-old native of Castle Hill, Carluke, Lanarkshire, Scotland who upon his death left a wife and five children, was buried in Coleman on Monday, June 22. Businesses in town closed for the funeral, which was one of the largest ever attended in Coleman.

Hughie Hunter, a native of Sauchie, Stillingshire, Scotland, left a widow and a son in Hillcrest.

Peter Ackers and Jack Sands, who had both worked at the McGillivray mine at Coleman a year previous, were both killed.

The explosion also took the life of William Trump, a native of Monmountshire, England, and well known to Coleman residents.

The tragedy reached much farther than the homes of Hillcrest and the Crowsnest Pass. The mines and mining families of Nova Scotia, who had provided much of the expertise at Hillcrest and who never stood far from the grief generated by their own mine accidents, bore the loss of twenty-two family members in the disaster. Westville's first coal mine opened in 1866, so its residents were well acquainted with mine fatalities, including those of Canada's first mine disaster. The town's newspaper of the day, the *Free Lance,* sketched out the lives of some of the Nova Scotians who were killed in the Hillcrest disaster, in an article reprinted in the *Bellevue Times* on July 2, 1914. William Neath, a man described as of clean life and splendid character, left a widow and a child. She took his remains with her back to Nova Scotia. Her brother, twenty-year-old Rod Wallace had gone west only a few months previous and was described as a fine young man. Wallace had intended to leave Hillcrest for good with Andrew, and head back to Nova Scotia. Both men were tired of mining and wanted to farm instead. His parents lost both a son and a son-in-law in one blow.

Another victim, thirty-one-year-old bachelor David Emery, had been in the West about a year and stood high in public esteem in his hometown of Westville. Forty-year-old Fred Bingham left a widow and four children. His remains were taken back to Westville, where his wife, who was also David Emery's sister, grieved doubly. James

Gray, who worked at Inverness and had left Westville only two months before, left a widow and four children at Vernon, B.C. He was described as a quiet, inoffensive man. Hillcrest superintendent James S. Quigley, a Scot, was recognized as smart, ambitious, and "a worker." "Jimmy" was highly esteemed by both management and men, and left a widow and five children. He had just completed a new house in Hillcrest and was about to move in. Thirty-five-year-old Tom Quigley, Jimmy's brother, worked at the Westville mines before he came west, and left a widow. John Hood and George Robertson, both from Stellarton, were also lost. The remains of James Barber were shipped back to Westville, and those of Angus Mackay and John A. McQuarrie were taken back to Inverness mines in Nova Scotia. In all, seven men returned in boxes to their homes in Nova Scotia where relatives and friends mourned their passing.

Among the victims was John B. McKinnon, a man renowned for his size and strength. At thirty years old, 6'4" and 215 pounds, he was known as the Samson of the Pass. The week before, at Frank, he had easily lifted a thirty-foot railway track, to the admiration of a crowd of miners. Among the last bodies to be recovered, he had been working on a pillar in mine no. 1. A bachelor, McKinnon had migrated west from Cape Breton.

Thomas Corkle had decided that June 19 would be his last shift at Hillcrest. In fact, he had only gone into the mine to remove his tools. Corkle planned a move to Nelson to work some metalliferous prospects near the town.

The wife and several children of James Bradshaw, one of the Hillcrest miners, were on their way to Hillcrest from Britain when the mine exploded. Mrs. Bradshaw received word that the mine had exploded upon landing at Quebec. The fate of her husband remained unknown as the family travelled from Quebec City to Montreal, where they were met by a family friend who informed them that James was missing in the mine. As they boarded their train for Hillcrest, they did not know that Bradshaw had already been found dead.

By Saturday morning, the day after the explosion, the outside world had become aware of the disaster. Alberta's lieutenant governor, George Bulyea, telegraphed Stirling to convey a message of sympathy from him and his wife to the relatives of sufferers and the people of Hillcrest, as well as a similar note that he had received from the government and people of Australia, which had also suffered coal mine disasters. King George sent a cablegram to the Governor General to express his sympathy with the families of those who perished. The British government sent a message of sympathy and offered assistance should it be needed. The Governor General of Canada and the Duke and Duchess of Connaught expressed their sympathy as well.

By midnight on Sunday, all but eight of the bodies had been recovered. There were forty-eight survivors listed. One hundred and eighty-nine were believed killed.

On Monday, June 22, while what seemed like an endless parade of funerals continued as mourners buried thirty-two more bodies, Superintendent Starnes wired an update on the events to RNWMP in Regina, who in turn informed Prime Minister Robert Borden. Borden indicated that he wanted as much information as possible about the extent of the disaster and the need for aid. The federal government dispatched an officer of the federal Department of Labour, J.B. McNiven, to Hillcrest to make a full investigation.

Just when it seemed that circumstances could not get any worse, a severe windstorm created havoc in the Pass, breaking CPR telegraph wires and the town's electrical wires, and blowing the roofs off buildings. The wind picked up the Union Hotel's brand new roof, which had sat on the ground awaiting installation, and tossed it on top of the CPR station's waiting room. The wind snapped off large trees, blew down electric poles, and, as a result, put nearly all of the telephones in Hillcrest and Bellevue out of commission. The cyclonic wind damaged homes as well, many of which were little more than shacks, some with canvas roofs that were easily torn and blown off. Some mining families, already prostrate with grief, suddenly found themselves without shelter.

On Tuesday, June 23, mourners buried twelve more bodies at Hillcrest.

In Fort MacLeod, Starnes was informed that drinking had again become an issue. Starnes in turn telegraphed the Alberta deputy attorney general that there were reports of a "good deal of drinking." Officials and clergymen, he wrote, have asked that the bar be closed again. Starnes concluded that it would be advisable to continue the suspension of the Union's liquor license for another couple of days. The government wired back its authorization, which would continue the suspension until Friday, June 26.

As if the families of dead miners did not have enough to deal with already, they soon faced another setback. Saturday was payday for the miners, but during the rescue and recovery effort, the issuance of cheques had been forgotten or pushed aside in favour of recovery efforts. Among those families who lived paycheque to paycheque, however, this quickly became a huge problem.

When word of the disaster reached nearby towns, relief efforts began, initiated for the most part by women. In Coleman, women made food in their homes, and the owners and staff of every restaurant and bakeshop began to prepare food. They carried it to the club room of the Institutional Church, and from there other volunteers drove it to Hillcrest. At Hillcrest, the Coleman women formed relief parties to assemble and distribute it to stricken families. A steady stream of vehicles drove for five days and nights between Coleman and Hillcrest with men for rescue teams, nurses, physicians, and food for rescuers and Hillcrest families alike. A small group of Hillcrest women took over the coal company office on Friday morning. There, they prepared wagonloads of food for the rescuers. Only on early Sunday morning did the women finally take a break.

Initially, the residents of the mining camps in the Crowsnest Pass, the City of Lethbridge, and the United Mine Workers of America met the serious need for relief. During the rescue and recovery, women had supplied sandwiches and buckets and washtubs of tea and coffee for the rescue and recovery teams, and for the police and miners in the wash house. By June 27, the women had formed an official relief committee, which also recruited two women from each town of Blairmore, Coleman, Frank, Bellevue, and Passburg.

Churches, through their female membership, led much of the relief efforts. The women of the Baptist Church canvassed the north side of Blairmore, and the Presbyterian Women's Guild the south side. This initial effort evolved into a larger fund for long-term relief authorized and partly funded by the provincial government. Once the provincial relief fund was in place, along with donations made by government, local organizations, and individuals, it distributed cash when and where it was needed to assist with groceries, clothing, and other living expenses.

On Tuesday, Lethbridge Mayor William Hardie arrived in Hillcrest to find out if he could help. He discovered that many of the miners' families were in a desperate situation, and immediately telephoned Lethbridge City Commissioner Arthur Reid to explain the circumstances. The City of Lethbridge donated two railcar loads of provisions, one full of flour and the other full of assorted groceries. Two mills donated fifty sacks of flour each, and the city guaranteed the remainder of the carload. This first material relief to Hillcrest had a value of about $1,000, plus a donation from the city of another $1,000 in cash.

On Wednesday, a relief store opened, stocked in large part by the railcar of supplies donated by Lethbridge. That morning, further relief arrived from Fernie, in the form of food and clothing. On the same day, the president of the CPR, Sir Thomas Shaughnessy telegraphed the head of the CPR's department of Natural Resources, P.L. Naismith, in Calgary to inform him that he had been authorized to put $2,000 towards the relief of grieving families.

The day after the explosion, Calgary's acting mayor, Michael Costello, telephoned Hillcrest Collieries manager Brown to find out whether Calgary could do anything to assist. The city had previously donated $5,000 to Fernie when the town had burned to the ground. In the end, the City of Calgary donated $2,500 towards the Hillcrest relief fund, and the T. Eaton Company donated $1,000. Every offer of aid was needed.

While miners and their families understood the urgent need for assistance, the general public, particularly beyond the Pass, did not. On Monday the 22nd, the MLA for Rocky Mountain, Robert E. Campbell, addressed Canadians via newspapers for aid for the victims' families. Virtually all of the newspapers in the Crowsnest Pass, Lethbridge, Calgary, and Edmonton encouraged their readers to contribute donations for the relief of the widows and orphans created by the disaster. The *Calgary Daily Herald, Lethbridge Daily Herald, Coleman Bulletin,* and *District Ledger* started their own relief funds. The Bank of Commerce, the Masonic Lodge (which lost twelve members of its Sentinel Lodge in the explosion), the Coleman Hotel, the Co-op Store, and the Salvation Army Band (which also lost a member) all donated to the fund. Campbell then appointed a relief committee composed of Judge Edward P. McNeill of the District Court of MacLeod, A.J. Carter, secretary of District 18 of the UMWA, and barrister Colin MacLeod, who would represent Hillcrest Collieries at the Commission of Inquiry. Contributions would be made to the Bellevue Union Bank, to the credit of the Hillcrest Widows' and Orphans' Relief Committee.

The Alberta government became aware that the relief effort needed organization and through an Order in Council established a *permanent* commission to handle the Hillcrest Relief Fund, and donated $20,000 to the cause. Judge McNeill was appointed chairman, with District Court Clerk W.C. Bryan as secretary, and the manager of the Union Bank at Hillcrest, Mr. Windsor , as treasurer. The establishment of this fund triggered the federal donation.

On June 26, the *Fernie Free Press* informed readers that the Dominion government had promised a donation of $50,000 for relief.

The newspaper also published another letter from MLA Campbell. This letter again appealed for assistance for the families of Hillcrest miners killed in the explosion. "The consequent suffering and destitution is beyond the realization of those of our people who have never witnesses or experienced the effects of such a disaster, but who nevertheless will sympathize with them in their unfortunate position."

Organizations such as the Masons and the UMWA, which contributed to the general fund, also created their own relief funds. On September 4, at a meeting of the union district, $525 was agreed upon as a donation for the benefit of union victims' dependents living outside Alberta. Another $4,000 was granted at the same meeting to the Hillcrest local for relief purposes. Perhaps the most important matter discussed at this meeting concerned the claims for compensation. Evidently union officials had discussed the possibility of an effort to collect damages on behalf of the widows and orphans, but decided that it would be impossible to succeed under liability or common law, and therefore moved that officials proceed to collect the compensation claims. Later in the month, the district granted another $1,500 for relief to Hillcrest, and a further $1,000 in October.

The RNWMP, meanwhile, had smaller but nonetheless contentious issues to deal with. On June 26, Superintendent Starnes reported to Regina that there was a slight difficulty between the miners' union and a representative of the Trust and Guarantee Company of Calgary as to the right of the trust company to look after the property of the dead men. Union officials objected to the involvement of the company on the grounds that it had been very slow in winding up the estates of the dead miners after the Bellevue explosion. As a result, Corporal Mead of the Bellevue detachment, acting under power of attorney for the widows, turned over only the personal effects that the agent of the Public Administrators (Trust and Guarantee Company) did not wish to bother about — rings, watches, et cetera. The secretary of the union local was present as these possessions were checked up and handed over to an agent of the Imperial Canadian Trust Company, as the union wished. Four hundred and fifty dollars found on the body of William Trump was handed over to the Trust and Guarantee Company.

Most of the estates of the men killed held a total value of less than $200, and in many cases much less than that. For instance, the value of the estate of Dan Kostyniuk totalled $18.71, of Steve Raitko, $6.61, of Chris Kosmik, $6.61, of Peter Fedoruk, $5.71, of John Tkaczuk, $16.31, and of Fred Ralnyk, $31.06.

The people of the Crowsnest Pass widely read the *Lethbridge Herald,* which was owned by Liberal MP William Buchanan, an old-style Laurier Liberal who supported low tariffs and reciprocity, or free trade.[2] In its June 26 issue, before the start of the Commission of Inquiry, the *Herald* published a small story with a large heading that read, "WIDOWS MAY NOT GET COMPENSATION. Startling Admission by One of Directors of the Hillcrest Collieries." The story intimated that one of the company's directors said that if the company went into liquidation, the bondholders would get first claim on the resources, meaning that there would be little chance of any compensation being paid to the miners' families. It was a warning from the company to miners, the union, widows, the provincial government, and the CPR that not only did the colliery need to resume operations soon, but also that lawsuits against the company could bring it down for good. Another story on the same page of the newspaper gave another reason to preserve the future of Hillcrest Collieries. In exactly the same size type, the headline read, "CPR TO ABANDON THE HOSMER MINE. Another Pass Town to be Wiped Off the Map — Mining is Too Expensive." The story reported that 1,200 men had lost their jobs when the CPR had shut down the Hosmer mine without warning. Workers had already begun to dismantle the machinery, and merchants and property owners were reported to be almost paralyzed with fear over the sudden loss in value of their land. Now District 18 of the UMWA had unemployed members with families to think about. Some of them would find work at Hillcrest, but not if the Commission of Inquiry pointed towards negligence by the company, which would initiate lawsuits for damages, and likely close the mine. These articles made everyone aware of what was at stake.

The stories in the *Herald* and the alarm it caused among miners and widows prompted Brown to call the newspaper on June 27 to deny its report. According to the *Herald,* Brown said, "The company expects to have no difficulty in meeting [the] drain the disaster will make on its resources." Brown completely denied that the company would be unable to pay compensation, as the *Herald*'s report claimed, and added "that none of the directors had ever authorized such a statement." The matter of the solvency of Hillcrest Collieries had been clarified, but Hosmer's problems had just begun.

Exactly one week after the explosion, the *Fernie Free Press* reported that riots had nearly broken out at Hosmer since the mine had been shut down so suddenly. Merchants demanded immediate payment of outstanding debts, and credit was cut off. That meant that miners had no money to live on until the balance of their wages was paid.

Now the Crowsnest Pass had a double crisis to deal with: the explosion at Hillcrest, which had left approximately ninety widows and more than 175 children fatherless, and the closure of the CPR's Hosmer mine, which left more than a thousand men without jobs.

At the same time, speculation about the cause of the disaster ran rampant. It had begun, naturally, on the day of the explosion, as miners, reporters, the general public, and grieving relatives tried to make some sense of the events. It continued as bodies were recovered, and became a part of the grieving process, a focus for anger, a means to find reason and order in chaos and meaningless death. It was a burst, a rock fall, a pick that struck a rock and created a spark, a match, a faulty lamp. It was human incompetence, a malignant company, a lazy miner. It was the revenge of God, the work of the Devil. But all of those affected directly knew or felt one thing for certain: pain, a pain of fathomless depth, of endless expanse, sharp and dull at the same time, throbbing and exhausting. And anyone who walked the streets of Hillcrest saw it in face after face: rage, numbness, longing, fear, confusion, regret, even guilt, and all of it glistening on cheeks as wooden coffins banged and slammed on the wagons as they trundled by, and then as miners shoveled the gravel that passed for soil in the Hillcrest cemetery onto the boxes that thumped hollowly until they could no longer be seen by those who stared down from the precipitous edges of the graves. Hollow. Hollow boxes, hollow souls, hollow lives. What could have caused such a mass extermination of human life?

On June 25, the Canadian representative of the Draeger Oxygen Apparatus Company of Pittsburgh arrived in Lethbridge. James Taylor had been a state mining inspector in Illinois for thirty years before he took up a position with Draeger, and had distinguished himself in connection with the Cherry, Illinois disaster in which he had saved many lives.[3] After he received word of the Hillcrest disaster, he hurried to the town and stayed there for several days. The Draeger Pulmotor, he learned, had saved the lives of several Hillcrest miners. While there, Taylor also gave a demonstration of a new Draeger self-rescuer to Lethbridge Mayor Hardie and a group of mining men.

A reporter asked Taylor for his thoughts on the cause of the disaster. Taylor made a comparison: just as a building settles if its foundations are removed, so too any mine sitting under a mountain. In Taylor's opinion, the settling caused a bump, which forced gas from pockets behind the coal with such velocity that it was propelled through the safety lamps, whose flames then ignited the gas. Taylor's explanation was plausible, but pure speculation.

In a letter written to a police official before the inquest began Inspector Junget said he believed that "[t]he cause of the explosion is unknown and it is doubtful that it will ever be known." This was certainly a curious remark for a policeman to make so soon after the disaster, even in an internal police letter, before he had heard testimony at the Commission of Inquiry or the Coroner's Inquest. The remark did, however, display Junget's insight, knowledge of men, and wisdom. He added, "The Hillcrest mine, which is considered a CPR mine, has always been considered one of the best and smoothest run mines in the pass, there having been less friction between the operators and miners than any place else. It was considered that it had been the best outfit of miners."

That evening Stirling wrote a long letter to Stocks in Edmonton, indicating that he had begun to formulate his own opinion as to the cause of the disaster. The letter detailed virtually all of the activity at the mine since the explosion. Stirling dismissed shot firing, faulty lamps, or detonators as the cause. As for caves, Stirling said the difficulty lay in the determination of whether the several caves found afterwards happened before the explosion, at the time of the explosion, or as a result of it. He said that if the Hillcrest coal dust were explosive, and the roof strikes sparked when it fell, because of the bitumen laminated into the stone he did not see how the Hillcrest mine could operate safely. Stirling also commented that it was particularly difficult to get information from the foreigners about their dependents and needs for relief and compensation purposes. With most of the bodies now out of the mine and buried, and cleanup and repair work underway in the mine, preparations had begun for the inquiry.

CHAPTER 11
Q & A

Circuit court judge Arthur A. Carpenter was the man appointed to head the Commission of Inquiry into the cause of the coal mine disaster that had been on the minds of almost everyone in Calgary since it hit the newspapers on the 19th of June. The government published notice of Carpenter's appointment in the *Alberta Gazette,* and suddenly, as newspapers got wind of the appointment, Carpenter himself became news.

Arthur Allan Carpenter, born in Hamilton, Ontario on September 3, 1873 earned his B.A. at the University of Toronto in 1894 and his LL.B. from Osgoode Hall in 1897, and was admitted to the Law Society of Upper Canada that same year. He practiced law in Hamilton for the next six years. But dissatisfied with what Central Canada had to offer him, Carpenter fixed his eyes on the far horizons of the West, and in 1903 boarded a train for the Northwest Territories. He set up his practice in Innisfail, near Red Deer, where he worked as the town solicitor until he was appointed to the district court at Fort MacLeod on November 21, 1907.

Meanwhile, Alberta had been admitted into Confederation as a province in 1905. Carpenter, subsequently, was accepted into the Law Society of Alberta as its 58th member in 1907, and moved to Calgary's district court in 1910. Tall, lean, self-confident, widely read, analytical, and not given to showing his emotions, Carpenter could provide the calm, steady hand needed in a situation in which emotions ran so high. He was virtually unknown in Hillcrest, except to members of the RNWMP from his time on the bench at Fort MacLeod, and likely by the counsel for the UMWA and Hillcrest Collieries. If there was a flaw in Carpenter's appointment, it might have been that he knew little about coal mining, but few judges at that time did. This single fact would later cause friction later with the UMWA.

On June 29, Superintendent Starnes arranged to hold the commission in the Sentinel Masonic Hall. Three bottles of blue ink were purchased: one for the judge, one for the RNWMP officer who would take his own notes of the proceedings, and one for the secretary who would take down the proceedings in shorthand. The stage was set for the first of the official inquiries.

A few days before the commission opened its hearings, the UMWA held their own investigation with a union-appointed committee. Its members, along with Chief Mining Inspector John Stirling, descended into the mine on the 29th to examine the scene of the disaster firsthand. With no official capacity, the union's investigation served to satisfy members that their interests were being looked after, to provide their own independent expert, Norman Fraser, with the information he would require to testify on the union's behalf, and to better inform union officials who would be present at the impending inquiry. The underground inspection, however, could not be completed on that particular day, as some parts of the mine were still inaccessible. Stirling, meanwhile, had already formulated his own analysis of the disaster and its cause.

The Commission of Inquiry commenced at 10 AM on July 2 with participants and spectators crowding the large second-floor

Masonic Hall. The representative for the Alberta government, William M. Campbell of MacLeod, had informed by telegram all those whose presence could be considered important. Colin MacLeod, a prosecutor from MacLeod, represented the owners of the mine. Robert Drinnan, the man previously considered for the position of general manager of Hillcrest Collieries, gave technical support. John R. Palmer, a Lethbridge barrister, assisted by Fraser, represented the interests of the UMWA. (A.J. Kappelle of Vancouver represented the Italian victims of the disaster, but he would not arrive for several days.) Also present were Robert Leavitt and James Burke, president and secretary of the Bellevue local, and Stirling. Judge Carpenter opened proceedings with a summary of the mandate he had been given, and reassured those present that anyone who wished to give evidence before the commission would be heard. Campbell, speaking for the crown, stated that every effort would be made to bring out any evidence that could throw any light on the cause of the disaster. Blueprints of the mine had been provided, and Stirling, he said, would be called to give the results of his investigations conducted inside the mine after the explosion.

Palmer, counsel for the union, then "made a formal protest against the form of the Commission, and submitted, with all due respect, that while the learned judge might be scrupulously impartial, nevertheless it was practically impossible for a layman to thoroughly understand the very technical and scientific evidence that will be given, without expert advice." Carpenter disagreed, but added that during the course of testimony he would further consider Palmer's objections.

Palmer then explained that the union's expert, Fraser, had been unable to make a full inspection of the mine because parts remained inaccessible. He therefore asked for an adjournment until Fraser completed his inspection and was able to compile his own report. In response, MacLeod, speaking for the mine's owners, pointed out that the entire mine was now accessible and Fraser could have made a quick inspection. He suggested to Carpenter that the proceedings continue, and the judge agreed. It was an odd situation. Stirling had not been forced to make a "quick inspection," and certainly Stirling had not been barred from accessing any part of the mine he wished to enter for his inspection. For Fraser to make a proper inspection would take not just hours, but days. One is left to wonder why the union left the matter so late.

The first order of business was Palmer's request that the company provide a list of documents, including an accounting of the number of men on each shift employed at the time of the disaster, a plan of the mine showing the present direction of ventilating currents, positions of stoppings, overcast, doors, et cetera, and the report books that the Mines Act required the management of the mine to keep on hand and up to date.

After these documents were provided and catalogued, Carpenter himself suggested an adjournment until the afternoon so the various counsels could draw up a plan of direction for the inquiry. The officials agreed.

At 2 PM, the lawyers first questioned one of the few men left alive with intimate knowledge of the mine: Hillcrest Collieries chief engineer William Hutchison.

Hutchison knew the inside of the Hillcrest mine intimately. He had walked every accessible foot of it, and was responsible with his brother David, the mine surveyor, for creation of the mine's blueprints. He was not a coal miner, however, nor was he an expert on ventilation, air flow, stoppings, safety lamps, or mining in general. As the first witness to testify before lawyers, officials and a judge who knew little about coal mining, the first questions asked of Hutchison were purely for instructional purposes, and focused on terminology and the interpretation of the blueprints.

On the morning of the explosion, Hutchison and his brother were about 150 feet down the old abandoned no. 3 slope. They were there to do a survey as part of a plan for future development work. Unexpectedly, the two men found themselves in a tremendous rush of air that had obviously broken through the wooden stoppings that were sealing off the air passage. Hutchison testified that his first impression was of a cyclone passing above. The rush of air brought with it small "shaley" rock and other debris, and no more than thirty seconds later, thick brownish-blue smoke. Within a minute, he said, they reached the outside, where he saw the same smoke pouring from the mouth of the no. 2 slant and a column of smoke from the bench above, somewhere near the no. 1 fan. The smoke from the no. 2 slant was coming out, he testified, with considerable force. Hutchison explained that the fan at the no. 2 entry had been stopped by the force of the air, but on Brown's orders, he and the hoist engineer restarted the fan in order to clear the smoke, so that men could get in or out of the mine.

At this point, Palmer began to ask Hutchison about the stoppage of the fan. It became clear that the fans had not been stopped in the past, except for servicing, and not for any great length of time. The stoppages had, he testified, been adequately supervised.

Palmer then asked Hutchison about water in the mine. Hutchison responded that the levels were very wet, as were the faces and entries. The slants were running with water and dripping constantly from the roof. There was so much water in the mine, he said, that pumps were on twenty to twenty-four hours a day.

On that first day the inquiry heard the testimony of another important witness, former Hillcrest fire boss Harry White, who, like Hutchison, had been involved in the rescue and recovery immediately after the explosion. White testified that he had been a fire boss at Hillcrest from 1908–1912, but at the time of the Hillcrest explosion was employed at Bellevue.[1] The current fire boss, Sam Charleton, killed in the disaster, was White's half-brother. White had first entered the mine at about 1 PM on the day of the disaster. On one of his rescue missions into the mine, he faced considerable smoke from fires burning in the chutes from rooms 24 to 30. Loose coal on the floor was burning, which produced

considerable smoke. White helped put out six of the fires by haul-
ing water for the firefighters.

MacLeod asked White about cave-ins. White said he had seen
signs of fresh caves. Several men had been buried by falls of coal,
but he noticed only one man struck by falling rock.

The lawyer asked White if he had been able to follow the direction and track of the explosion. White said that he had, and that
he believed the explosion had started somewhere around room
33. That was where he and Bob Creadon had found the bodies of
his brother-in-law, George Hicken, and Charleton, among others.

Palmer returned to the topic of dust. White said that he would
not call Hillcrest a dusty mine, but that during his rescue efforts,
he had seen a lot of dust in the rooms where the explosion had
been most violent. White said, in his opinion, there had been a
lot of dust in these rooms before the explosion, too.

MacLeod and asked him about his discovery of Hicken's body.
White found Hicken at the second crosscut at room 33, which was
where White believed the explosion had occurred.

Campbell then asked White about rock that had fallen in the
old workings at Bellevue, which, when it fell, created sparks.
White said this was true, and that he had seen it happen in the
Hillcrest mine as well. But before White could explain, the commission broke for the day.

The investigation resumed on July 3, but Palmer again addressed Carpenter's lack of expertise with respect to coal mining.
Palmer said that the union had fully expected the government
to appoint an expert to sit with the judge. Carpenter replied that
he had no power to appoint anyone like that, but could call any
expert evidence that might be needed.

Harry White was again called to give testimony. He said that
he had seen rock fall and strike sparks during a previous night
examination at Hillcrest. White reported the cave, and had discussed the sparks he saw with the superintendent at the time,
James Quigley.

Next to testify was the man responsible for the mechanical operation of the fans. Master mechanic Thomas Hargreaves swore that both fans had been operating at the time of the explosion. He added that since he began working at Hillcrest on June 1, no fan had stopped due to mechanical failure.

Hargreaves' testimony is more important not for what he was asked, but for what he wasn't asked. Curiously, he was never asked about the stoppage of the no. 1 fan immediately after the explosion. No questions were asked about the nature of the damage to the fan, or about why it had taken half an hour to repair. No one called the assistant outside engineer to ask why he had been at the no. 1 fan just before the explosion, whether it was a routine inspection, or, if not, what he had repaired.

On the afternoon of the 3rd, White was recalled as a witness and asked about his discovery of the body of fire boss Sam Charleton. White explained that fire bosses always carried an electric battery, a key to the battery, and a cable for firing shots. When he found Charleton, his cable was wrapped around his body, suggesting it was not in use. If he had been in the act of firing a shot, White said, it would have been strung along the working. White's testimony effectively ruled out shot firing as a direct cause of the explosion.

The last important witness of the day was fire boss William Adlam, the man who had examined the mine prior to the morning shift on the day of the explosion, and whose brother Herbert was killed in the disaster. Adlam, with twenty-three years' experience coal mining in England and Canada, testified that he had found gas in seven rooms — no more than usual — and that the ventilation was good. As for dust, Adlam said that the amount was not out of the ordinary, and it was not what he would call a dusty mine.

Adlam said he had seen rock fall in the mine before, but no sparks created as a result. He said that he had never heard anyone connected with the mine mention falling rock creating sparks either.

Immediately after he left the mine that morning, Adlam had posted and reported his findings of gas. It was the job of the next fire bosses on shift to send brattice men in to remove the gas. Normally, they went in about twenty minutes before the miners.

Palmer observed that in Adlam's reports back to May, he had frequently reported gas in the same places. Adlam agreed that this was the case. When he found gas, he would fence off the area with boards and planks, and a sign that said, "No Road."

Adlam had not gone to bed when he got off shift that morning. He was in his backyard when he heard the engine house whistle blow the alarm of three blasts. Immediately he put on his coat and rushed up to the mine to help deal with what he assumed was a fire in one of the outside buildings. When he got there, he saw smoke pouring from the no. 1 fan. He wasn't able to enter the mine until about three in the afternoon because of the heavy smoke, but stayed inside until 2 AM on Saturday repairing ventilation. At 9:15 he entered again with a recovery crew to look for bodies. They found fifteen. Some of the men appeared to have moved from the face in an effort to escape, but had succumbed to carbon monoxide.

Adlam agreed when questioned that the explosion had not happened in the section of the mine off the no. 1 slant, and that the direction of force there indicated that the explosion occurred in the lower workings.

Fire boss Dan Briscoe testified on Friday, July 4 that he left the mine at 10:30 PM on June 18, the day before the explosion. He observed that the amount of dust appeared normal, that it was not a dusty mine, and that there was lots of moisture present. He said the ventilation was efficient, and there had been no problems with the fans. The mine, he said, was run carefully, and he had never seen a cave occur in the mine, nor had he seen reports or heard anyone remark that the rock produced sparks when it did fall.

During the questions asked by the counsel for the UMWA, it became apparent to counsel for the company that Palmer was trying to discredit the company rather than determine the cause of the explosion, which led MacLeod to finally complain to Judge Carpenter. Palmer denied that he was behaving improperly, but soon aroused the judge's concern again when he tried to introduce hearsay evidence. Carpenter said he would not have it, but Palmer again denied any impropriety and pressed on.

The commissioners next questioned the fire boss who had survived, John Ironmonger. He had worked at the mine for eighteen months previously, and had five relatives in the employ of the company: two brothers, a cousin, and two that were married to his cousins. All were killed. Ironmonger had been in No. 1 North level when the mine exploded. He began to run for the exit, and said that he was about 500 or 600 feet from the entry to Slant 2 when he met the smoke. Ironmonger said that he had never heard anyone ever mention that sparks could be caused by rock when it fell from the roof in the mine, and had not seen sparks himself when he had witnessed caves. He also said, when questioned, that he had seen what was known as "travelling gas" in the No. 2 South level, similar to Briscoe's testimony that he had seen travelling gas. Travelling gas travels with the air current, as opposed to a pocket of gas which sits still in a pocket at the roof. Palmer began to ask a lengthy series of questions of Ironmonger about how many men Ironmonger had seen in particular places in the mine, and their respective jobs, but again, MacLeod objected that these were calculated to find fault with the company rather than to ascertain the cause of the disaster. MacLeod said it appeared that "the object of my learned friend is to throw disparagement on the management, or something the management did inconsistent with the Act. If the company have done anything wrong there is the procedure to take them into the Courts, but this enquiry is to ascertain the cause of this explosion." Palmer was obstinate on this point, and continued his line of questioning.

He asked Ironmonger about Quigley, the mine superintendent, with whom he had entered the mine. The two men had become separated as they went about their separate duties. Ironmonger admitted that it seemed likely that after the explosion, uninjured by the force of the blast, Quigley had led four men along the level from chute 74 in a hunt for a way out when they had been over-come by carbon monoxide, and died. Ironmonger's testimony ended with an adjournment until Monday morning.

The July 4 edition of the *District Ledger*, the official organ of the UMWA, carried extensive front-page coverage of the activity on the first day of the Hillcrest inquiry. It also published a story in small type, next to the page-one report, about a different matter altogether, but one that would have a profound effect upon not only the lives of people in the Pass, but also upon the rest of the civilized world.

A man had shot dead the heir to the Austro-Hungarian throne, Archduke Franz Ferdinand and his morganatic wife Sophie, the Duchess of Hohenberg, in the main street of Sarajevo, Bosnia on the June 28, 1914. The murders triggered the conflagration that would become the Great War, and swept the Hillcrest explosion from the minds of most Canadians.

As the inquiry continued in the town, Hillcrest miners had undertaken repair work and prepared to resume mine operations. A number of key positions had to be filled, however, and on July 6, the secretary treasurer of Hillcrest Collieries informed Alberta's minister of public works that Robert T. Stewart of Fernie had been appointed mine manager to replace the deceased Quigley.

The inquiry resumed on Monday, July 6, with a tour of the mine. Judge Carpenter, Campbell, Aspinall, Palmer, Fraser, Bren-nan, Constable Hancock, and a fire boss entered the mine to examine the various places where the bodies were found, and to view the mine for a better appreciation of what happened and where. The party, led by Aspinall, entered into the rock tunnel no. 1 entry, and after two hours, exited the no. 2 slope. On the surface, they inspected the hoist buildings and the no. 1 and no. 2 fans. Their tour complete, the inquiry resumed in the Masonic hall.

This tour was one of many the officials conducted to investigate the cause of the explosion. On one of these inspections, the group was ascending a room off no. 2 south when fire boss stopped and held up his hand. As Hutchison tells it, "I stopped also, knowing that his signal meant danger." Chief Mine Inspector Stirling barged forward, his manner patronizing: "What's the matter, getting scared, are you?" At that moment, a large section of roof caved just a short distance ahead — a cave that would have killed them, had the fire boss not stopped them.

In the afternoon, fire boss Walter Rose testified first. Rose said that he had seen a cave cause sparks in the Hillcrest mine about four years previous, but had not heard anyone report or speak about sparks since.

Electrician Andrew Wilson testified that he inspected the electrical wiring for the lights and pumps daily. On the day before the explosion, he had made some repairs to the lead sheath on the cable that supplied power to the pumps. It had been damaged by runaway trips, but had been immediately repaired. After the explosion, he said, the no. 1 fan had been out of commission for fully half an hour. It was restarted at about 10 AM.

Thomas Brown, the no. 2 hoist man, testified that the hoist fan had not to his knowledge ever stopped due to mechanical breakdown, and was running at the time of the explosion.

Ed Keith, the power house engineer, testified that the fan had been running up until the time he went off shift at 7 AM. It had been deliberately shut down about two months before the accident to adjust its bearings, but only for about ten minutes.

Dr. Allan Ross testified in response to questions about causes of death, but was only able to read from a list made by other physicians. Ross had viewed bodies not for the purpose of determining their cause of death, or to make a postmortem examination, but simply to pass them for burial. In fact, he said, the list of causes of death made no mention of carbon monoxide poisoning. Ross suggested that an ignorant stenographer who prepared the list was responsible for the omissions. Palmer asked him if the causes of death could be obtained from the death certificates, but Ross informed him that all that is marked down on them is "Killed in mine explosion." Ross had, however, observed that a great many had obviously died from carbon monoxide poisoning — obvious because carbon monoxide poisoning gave a distinctive pinkish cast to the victims' skin. Of the men he had seen dead he believed that twenty-three had died from carbon monoxide poisoning.

That afternoon, while the commission sat, miners recovered another body from the mine. At about 3 PM, a recovery team found the remains of Joseph Oakley buried under a pile of coal and rock. The family took the body to Sparwood, B.C. to be buried.

The next morning, July 7, John Brown, who had managed the Hillcrest mine for four years, testified. First, Campbell asked if Brown knew "how the law was carried out with regard to the searching of miners or other men going into the mine, for matches?" Brown answered that he had heard the fire boss, pit boss, and manager talk about searching the men. Campbell asked Brown if he thought the law was enforced, and Brown answered yes. He said he did not know of any occasions when matches were found on any of the men going into the mine.

Brown evaded the issue of whether they had actually conducted a search or whether he heard them say that they had conducted a search. What was not asked was, "Was a search conducted, and when?" The manner of the question, the evasiveness of the answer, and Palmer's failure to press the line of questioning serves as a perfect illustration of the desire by all parties to evade difficult questions, as well as answers that might point to individual responsibility. Certainly, the union wanted no suggestion that a miner might have been responsible for the disaster, and the company wanted no suggestion that management had failed to conduct searches for potentially dangerous matches and smoking materials, searches that were vigourously protested by miners as unnecessary, and were certain to generate hostility between management and union.

Brown later said that he, too, had never heard of sparks caused by falling rock.

Kappelle, who had been hired by the Italian consulate in Vancouver because so many Italians had been killed in the explosion, did not pull his punches. He asked Brown to repeat that no precaution had been taken to eliminate dust. Brown answered, "That is right."

"No precaution whatever?"

"No."

"Why not?"

"Because there was not sufficient dust there to take precautions."

Kappelle then asked Brown if he had found dust in the mine since the explosion. Brown replied that he had found considerable amounts in certain places.

Kappelle asked if it would be a fair assumption that there must have been a large quantity of gas gathered throughout the workings of the mine. Brown answered in the affirmative.

"And that being the case, can you say to this court that the mine was properly ventilated?"

"Yes sir, I can say to the court the mine was properly ventilated."

"Notwithstanding the fact that all this gas was accumulated in the different workings?"

"There was no accumulation of gas in the mine at the day of the explosion according to the reports I have seen."

Kappelle then asked Brown if he thought it was possible for the gas that caused the explosion to have accumulated so quickly after the fire boss' report the morning of the explosion, the report that concluded there was not an unusual amount of gas in the mine. "That is a question I am not prepared to answer." Brown's reply no doubt echoed in the mind of everyone present. How could the mine have exploded if coal dust was not a problem, if fire boss reports indicated that neither dust nor gas were problems, and if there were adequate ventilation?

For Brown, to answer MacLeod's questions instead of those asked by Kappelle would have come as a relief, as they were apparently an effort to clarify lines of responsibility, and shift it away from Brown. Brown, according to his own testimony, had no special jurisdiction in the mine. The mine's manager, Quigley, whose certificate was registered with the department, was the man in charge of its general workings. Quigley was the expert. Quigley was the man to whom the department would look, and upon whom the department would fix responsibility for anything that occurred in the mine. One is left to wonder why Brown had been hired, at great expense, to run the mine. And now Quigley was dead.

MacLeod asked whether dust had been watered at the Michel, B.C. mine, which was far dustier than Hillcrest. Brown replied that water was not used there, or at any other mine in Alberta or southern B.C.

MacLeod asked if Brown had ever received complaints about the mine's ventilation. Brown responded, "No." Only once had a mines inspector complained about ventilation, and that issue had been rectified with the installation of the new fan at no. 1.

At 3:15 PM, the commission adjourned to revisit the mine. Judge Carpenter, Campbell, Stirling, J.C. Roberts (a mining expert with the U.S. Department of the Interior), and Constable Hancock spent about two hours in the mine, which just that morning had resumed operations for the first time since the explosion. The eighty men per shift produced 133 tons of coal.

However, Stirling, who had remained silent about the resumption of operations, inexplicably telegraphed Hillcrest Collieries eleven days later, on July 18, saying that he strongly disapproved of the mine being reopened until a definite understanding could be reached about future mining operations. He said he would soon arrive in Frank for a thorough discussion of the matter.

Managing Director J.M. Mackie, still at Hillcrest on the 18th, responded immediately by letter. He expressed his surprise at Stirling's claim that he did not know they were working, as he had been in the mine and seen the men at work himself. Mackie also pointed out that District Inspector Aspinall was aware no later than the 10th that the mine was back in operation.

George Wilde, who said he had worked in perhaps a dozen different mines in his twenty-six-year career, was the first to testify on the morning of July 8. He said that he had never had cause to complain to anyone about ventilation in the mine. Wilde said that he had never heard of the presence of gas that was not covered in the fire boss reports, and that he believed those reports were accurate. He said most of the places he worked in were not dusty. Wilde said he had never seen rock fall from a roof, but anyone who struck rock with a pick was likely to create sparks.

Kappelle then asked Wilde, "Were you in the cemetery where the bodies were buried?"

"Yes."

"Didn't you see any dust down there?"

"Not to my knowledge."

"The reason I asked you that-you spoke to me yesterday — don't you remember speaking to me yesterday, introducing yourself as George Wilde?

"Yes."

"Will you think again, please? Did you find ashes or other dust down at the cemetery where these bodies are buried when you went down there?"

"No."

Wilde's responses were falsehoods, but he was not challenged. Dust had settled on the ground at the cemetery. In fact, coal dust covered the entire town. Women who hung their wash on clotheslines outside often complained about the dust settling on their clean washes. Wilde denied Kappelle the point, but it was irrelevant anyway with respect to the cause of the explosion. The dust was generated by the tipple, not by the mine. The question presents itself: why would Wilde lie when all he had to say was that the dust came from the tipple.

District Inspector Andrew Scott was sworn in. His report from April 2, 1914 was presented as evidence. At the time Scott found the ventilation and timber good, and found no explosive gas. He told Palmer that he had been into the mine over the last few days and had found much more dust than he had in April.

Thirty-year-old Sam Wallace had worked in mines since he was twelve. He said there was no undue dust in the mine, and he had never experienced poor ventilation at Hillcrest. Kappelle took over the questioning, and asked again about dust. Wallace's job had been to shovel loose coal dust into banks or ridges, and set the wooden stoppings on top. Wallace said there was perhaps six inches of dust on the floor.

The inquiry called James Gorton, a miner with four years' experience at Hillcrest and secretary of the union local, who was not in the mine at the time of the explosion. He testified about a grievance that the union local had made against the company. The company wanted to cut back on the firing of shots, because they claimed it broke up the coal, and wanted the miners to get it out with a pick instead. The miners, however, said they would not stand for it, as the extraction of coal with picks was far slower than extraction with shots, and the price they got on the tonnage would not let them make a living. The miners had won their point and the shooting had resumed.

Gorton was also part of a pit committee, along with F. Pearson and George Pounder, that had inspected the mine on May 18. The committee conducted mine examinations by union agreement from time to time on behalf of the men, always accompanied by the superintendent. They found the ventilation to be good. Only two rooms had enough gas in them to put the light out. Dust was not found to be a problem.

On the afternoon of the 8th, Frank Aspinall, former inspector for the Crowsnest Pass, now inspector for the Calgary district, testified. In an inspection report from July 4, 1913, Aspinall reported considerable dust at the face of no. 1 south entry, in the counter entry, and in multiple rooms. Another report from July 1912 complained about ventilation. He had found dangerous quantities of gas and the brattices to be completely inadequate. When he mentioned the problem to Quigley, the superintendent said there had been a shortage of lumber. He also offered assurances that a new fan had been shipped and was on its way. Aspinall told Quigley that the mine ought to have been shut down rather than allowed to operate in dangerous conditions.

Palmer, representing the miner's union, asked Aspinall, "Have you been in the mine quite recently?"

"Yes."

"And been all over it. And how would you say it compared now with its condition during your inspectorship in this district, as to dust? Has it improved or not?"

"No, it has gone worse."

"And at the time you were inspector in this district, would you have called this a dusty mine?"

"Yes, undoubtedly."

Palmer then asked about ventilation, and in particular the management's change of the no. 2 fan from an exhaust fan to a force fan. Aspinall said he considered the change to be seriously flawed. Palmer asked what Aspinall would have done if he had been inspector and found the ventilation altered in this manner.

Aspinall's reply was to the point: "Shut the mine down or prosecute the company if it was not changed."

Palmer then asked about the number of men who worked under particular fire bosses. "Fire bosses have come up here and told us it would not be possible for them to give us any idea of the number of men working for them from time to time in their district. What do you think about that?"

"Do you want me to express my opinion as to whether the fire bosses are telling the truth or not, is that it?"

"Yes."

"Well I think they are lying."

"You think they are lying?"

"I am sure of it."

Palmer asked Aspinall what he thought caused the explosion. Aspinall replied that gas was the only possible cause. Only a blown-out shot could have ignited coal dust, but a blown-out shot had been ruled out.

The issue then became: what had ignited the gas? Aspinall ruled out a defective lamp, as too many things would have had to go wrong for this to happen: the miner would had to have been given a defective lamp, the fire boss and miner would both have had to examine it and pass it in a defective condition, and that particular lamp would have to have been in the exact place where there was an explosive mixture of gas.

A cave that set off sparks was also eliminated as a potential cause. The caves in the area where the explosion had originated had been caused by the blast itself.

MacLeod, the company lawyer, then cross-examined Aspinall. His questions suggested that the explosion did not originate at or near where Adlam had reported an accumulation of gas on the morning of the explosion, because in that area there was little evidence of strong force.

Aspinall disagreed. "The way I figure it, the great probabilities are that where the explosion originated there is very little evidence of force."

Aspinall's powerful testimony was the last for the day. Radically at odds with the picture presented by other witnesses, it must have been unsettling for the company and the union to hear.

Aspinall had not allowed himself to be intimidated or manipulated by his questioners, and had given direct, unequivocal answers that lacked the evasiveness of Brown's testimony.

The next morning, July 9, fire boss John Ironmonger was recalled. He testified that he had examined the lamps of the brattice men who went into the mine that morning. But he was not asked, during his initial testimony or his recall, at what time the brattice men had gone into the mine to remove gas. It was a key point, and extraordinary that it had been overlooked by the lawyers.

Stirling took the stand and read his notes taken during his inspection of the mine after the disaster. The report starkly described the effects of the explosion, and the tremendous force that had been unleashed in the dark corridors of the mine on June 19. He found that a compressed air receiver had been blown for a distance of 250 feet towards the no. 1 slope on the no. 1 south level.

Norman Fraser, the expert retained by the union, was sworn in. Palmer, the union's lawyer, asked him if he would call Hillcrest a dusty mine. Fraser responded that yes, he would call it a dusty mine, but not a gassy mine by the standards in the Crowsnest Pass.

Fraser concluded that the explosion resulted from the ignition of gas, and was propagated by coal dust. He could not say, however, how the gas was ignited. Two shot holes were found in the face at no. 32. Both were properly placed and ready for firing, but had not been fired. Fire boss Sam Charleton was found dead at no. 35, which Fraser said would lead one to conclude that Charleton did not fire them because he did not think it was safe to do so. Fraser added that "there would be clouds of fine coal dust in the air caused by the coal running down the chutes and as gas would be coming from the working places, all the elements for an explosion would be present, and it would only require the initial flame to start the explosion."

Fraser went further. "I am convinced," he said, "that had there been a current of fresh air supplied to the faces of no. 2 south to dilute and render harmless noxious gas as required by section 58

of the Mines Act this accident would not have happened." Fraser
had given explosive testimony, and in effect, bait for the company
counsel MacLeod, who accused him of exaggeration and of using
the word "approximately" to cover his dishonesty in giving an
interpretation biased towards the company. MacLeod challenged
Fraser on his contention that Hillcrest was a dusty mine, contrary
to the testimony of Hillcrest miners, Gorton, White, Wallace, and
Wilde, as well as the other fire bosses. In MacLeod's words, "This
is a hostile witness and he is exaggerating."

MacLeod then began to ask Fraser where he thought the seat of
the explosion might have been. Fraser's answered that he thought
it occurred somewhere between rooms 31 to 47 off the no. 2 south
level. Some time was spent on room 42, the crosscut off the face
of 42, and the body of the man found there who died from afterdamp. MacLeod focused his questions on the direction of force,
the amount of force observed by Fraser, and whether there were
signs of coking and burning. Then MacLeod asked specifically
about the direction of force at the crosscut between 42 and 41.
For the first time in his testimony, Fraser hesitated.

"Well, above that I cannot tell," he said. "I cannot decide, but
I saw something; you might think it was going down." One has
the sense that Fraser had an instinct that the seat of the explosion was near here, but he could not articulate it, and in a formal
setting where legal rules of evidence applied, his instincts were
all but worthless.

When the inquiry resumed sitting on the morning of July 10,
MacLeod resumed his questioning of Fraser. He began by using
Fraser to establish that the company was not negligent in its
choice not to apply water to dust in the mine. Fraser agreed. He
said that when he had worked at Michel, parts of that mine were
far dustier than at Hillcrest, and he had not watered dust there;
he had merely discontinued shot firing in the dusty areas.

Later in the morning, the commission recalled George Wilde
to the stand. He testified to his recovery efforts in the rooms off

the no. 1 slant, and his discovery of bodies. He and men from Passburg and Michel had found four bodies in room 57, and carried them down the room to an old travelling way along the old south level. Wilde's claim that they had moved the bodies contradicted Adlam's earlier testimony that it appeared the men had moved on their own. In fact, the bodies had been moved by Wilde's recovery team. The new and contradictory evidence disturbed Judge Carpenter. He reminded counsel for all sides that he had stated at the very beginning of the inquiry that he wanted all the evidence as to facts in *before* the expert witnesses testified. Carpenter's rebuke prompted Palmer to defend himself. He said that he had planned to call rescue teams to testify, which would have clarified such matters, but Campbell had argued that it was unnecessary, and MacLeod had responded that he would not press the matter. The hostility between MacLeod and Palmer suddenly broke out into the open.

"My learned friend has offered this Commission no assistance at all," said MacLeod.

Palmer replied, "It was my object in the first instance to produce evidence showing where the bodies were located and my learned friend urged upon Your Honour he did not see the necessity for that and did not press it."

MacLeod seemed to be making excuses at this point. "I suggested as a matter of fact it would assist the Commission if the men were located and if the men moved it would be evidence of what they died of and the force of the explosion and to that end I asked the questions of the fire bosses as to the rescue. The only question I raised or remark I made was because Your Honour questioned the reliability of Dr. Ross's evidence, inasmuch as most of it was hearsay, and I said I was not pressing that point."

MacLeod now was babbling. "That had nothing to do with the evidence that is now being submitted by this witness," Carpenter said.

MacLeod continued his evasion. "No, the doctor was giving evidence for the other doctors and Your Honour said it was not good evidence and I said I would not press it."

Carpenter took control of the proceedings. "I must say it seems to me that this evidence is of importance and it is not too late yet to put it in. This is a Commission of Enquiry and not a civil action." Carpenter was referring to the evidence about the movements of men and bodies in the mine both before and after the explosion.

At that point, MacLeod resumed questioning Wilde, who testified that the bodies in room 32 were badly mutilated and had felt the force of the explosion considerably, whereas in 57 that was not the case. The four men he found in 57 were all close together at their working place where they had been skipping the pillars. Wilde testified again that he and those with him had taken the bodies out to a place where the stretcher bearers would take them to the surface. Wilde then testified that Adlam, the fire boss, had been with them, but had left to try to put out fires down below that he said were not safe. Wilde could not remember if Adlam had left before or after they removed the bodies. Adlam *was* in 57 with Wilde when they had found the bodies. In fact, *he had helped to identify them.* Wilde's testimony was startling for its illumination of the weakness of Adlam's testimony — which in this instance Wilde revealed to be altogether false.

Adlam was recalled to the witness stand, but the lawyers seemed to have developed a sudden amnesia with respect to his earlier testimony, which Wilde's account appeared to invalidate. They asked him nothing about the discovery of the bodies, nor whether they had walked on their own after the explosion or had been carried when dead. Instead, Adlam was asked about his examination of the mine prior to the explosion. In particular, he was asked about the amounts of gas he found, which were, he now admitted, considerable. The worst was in room 43, which put out his lamp and which he then fenced off.

Adlam testified that he did not know if the brattice men went in at their usual time of 6:30 AM, half an hour ahead of the shift. This statement was key, because Adlam testified that, as a rule, the brattice men would not go in until he came out and left copies of his report: one posted on the board, and one on the table for the fire bosses. On the 19th he came out at 6:20 AM, and left the report. The next shift of fire bosses would be there by 6:30, and were supposed to give instructions to the brattice men, who would then enter the mine to clear out the gas. They cannot enter the mine, Adlam testified, until the morning fire bosses check their lamps, and get their instructions as to the locations of gas.

The commission called Robert G. Drinnan to testify. He now lived in Edmonton, and had more than twenty years of coal mining experience as an apprentice mining engineer, mine surveyor, fire boss, under manager, and manager. He had worked as a mine examiner in B.C. and Alberta, had been manager of the Crow's Nest Pass Coal Company at Fernie for seven years, and had earned certificates as mine manager in Alberta, B.C., and Great Britain. Drinnan had been involved in the investigation into the Bellevue explosion in 1910, and had examined the Hillcrest mine after the explosion to try to find the cause.

During the questioning, Drinnan referred to "the old south level" as an area that he had examined. The term confused Carpenter, who asked Drinnan to refer to the place differently. However, the lawyer for the company, MacLeod, ignored Carpenter's request and continued to use Drinnan's term.

Frustrated, Carpenter pointed to the blueprint and asked, "This is what you mean by the old south level here, is it?" He received no reply. The refusal to accede to Judge Carpenter's request for clarity can only be interpreted as an expression of hostility or contempt for the judge on MacLeod's part.

Drinnan absolved the company of any responsibility. MacLeod asked, "Are you of the opinion that the current of fresh air that went into no. 2 mine, if no. 2 fan was exhausting, would it be anything to do with this accident?"

"No, I do not think it would have any bearing on the accident at all," he replied.

MacLeod then asked about the production of carbon monoxide and carbon dioxide by the explosion. Drinnan responded that a dust explosion tends to produce a great deal of carbon monoxide, whereas in a gas explosion you would expect more complete combustion and therefore the production of more carbon dioxide. Drinnan agreed that the fact that Hutchison and the fire bosses were able to get to the face of no. 2 south quickly during their rescue efforts indicated that coal dust played a very small part in the explosion.

Drinnan could only get away with such a speculative answer because no evidence had been forthcoming about the causes of death. The medical doctor called as a witness had not examined the bodies for cause of death; his job was only to certify that the dead bodies were dead so that they could be buried. No physician was called to testify who could speak about causes of death. Consequently, not even a rough estimate of the causes of death was available. Yet company documents show that most causes of death had been attributed.

Drinnan further agreed that the only danger in the mine would be from firing shots in a dusty area, and since shots had not been fired immediately prior to the explosion, the mine was safe. Therefore one could conclude an absence of negligence on the company's part. MacLeod asked, "There is nothing that the company might have done, that you could see, to have guarded against this accident?"

Drinnan answered, "No, nothing that I could see."

Drinnan's conclusion was a large leap by any standard, one that completely ignored the fact that most of the men killed had succumbed to afterdamp, or carbon monoxide. Palmer took over the questioning with that in mind. He asked Drinnan if he would quarrel with Fraser when he said that it would be possible that the ventilating system of a mine would be responsible for an explosion. Drinnan responded that the ventilation in any mine would.

Sterling testified last. It was MacLeod who summed up what had been said to that point: "Now, as a matter of fact, as far as this explosion is concerned, there must have been something happened in that mine of which we have no evidence — something unforeseen?

"Yes," replied Sterling.

"That is to say, although we have no evidence of the fact, it might have been that through some cause or other — the ventilation system might have been disarranged, which might cause an accumulation of gas — of fire damp — and just at that moment a defective safety lamp, a miner with a match, or anything — something of that nature — must have occurred?"

"It is possible, of course; it would be a coincidence."

The same day, July 10, J.C. Roberts of the U.S. Department of the Interior, Bureau of Mines Experiment Station, an expert who had spent considerable time in Hillcrest after the explosion, wrote a letter to Stirling. He wrote, "I have the honour to inform you that after nearly three weeks of continuous investigation and search regarding the Hillcrest Collieries disaster, I have failed to arrive at any definite conclusion as to the initial cause of the explosion. The bulk of my investigations were made in your company together with that of your district inspectors and officials of the Hillcrest Collieries Ltd."

The next day, Judge Carpenter asked counsel for concluding remarks. Afterwards, he said that he would render his report in due time. This concluded the inquiry.

At about 12:30 PM on July 14, a recovery team found the last body that would be recovered from the Hillcrest disaster. The team had believed that William G. Miller had already been buried. The body was badly burned on the head and shoulders, and his head was smashed, but the identification was positive, and Miller was shortly thereafter buried at Hillcrest. The only body that remained in the mine, apparently, was that of Sydney Bainbridge, who was still buried under a fall of rock, and considered too dangerous to recover.

CHAPTER 12
THE ASHWORTH LETTERS

At the time of the Hillcrest explosion, a consulting mining engineer named James Ashworth resided in Vancouver, British Columbia. A member of the South Wales Institute of Engineers, the Vancouver Chamber of Mines, the Canadian Mining Institute, and the General Manager of Crow's Nest Pass Coal Company from 1909–1911, Ashworth had earned a reputation as a mining expert. His particular interest lay in the field of mine explosions. He co-wrote a paper with his brother on the causes of coal dust explosions for the journal *Canadian Institute of Mining*, and closely followed accounts and reports of explosions wherever they occurred. He had also written papers on the cause of the Killingworth explosion in New South Wales, Australia, and on the Welsh Senghenydd colliery disaster of October 14, 1913, both published in mine engineering journals in Britain. The Senghenydd disaster, Britain's worst, killed 489 miners at a mine near Caerphill, Glamorgan, in Wales. At first, the usual suspects were blamed for the explosion — sparks from falling rock or electrical equipment

— but after the proper questions were asked and answered, the Royal Commission determined that a human action had caused the disaster: an open light. This commission was made up of mining men, not lawyers.

Ashworth had personally testified at the inquest into the Bellevue disaster and was therefore acquainted with the provincial solicitor and the other men involved. Shortly after the Hillcrest disaster, Ashworth began to write letters, eleven in total, to various officials connected with it. Although one could question Ashworth's motives for entering into this correspondence, they have no bearing on the value of his observations, opinions, and comments in his letters. It is clear he sought employment as a consultant in the disaster. In his letter of July 24, he offered his services to the minister of public works to help find the cause of the disaster. Then, well after his services as a consultant would have been useful, he commented in a letter to Stirling that he suspected coal mining was in better shape in Alberta than it was in B.C., and that a job as an inspector there "has attractions." He sought employment, but he was also very experienced with respect to the issue of mine explosions and their causes.

Ashworth received replies from Stirling and two government ministers. One additional, and unusual, letter is from Ashworth's young daughter, who apparently wrote to Stirling without her father's knowledge, begging Stirling to let her father investigate the explosion or use him as a resource in the investigation. It was a childish letter, and could be dismissed as such.

Ashworth's own letters beg for detailed information about the explosion, and the information he sought was passed on to him in a polite but distant manner that implied Alberta mining officials had no intention of making use of his knowledge, nor any interest in whatever conclusions he might reach about the cause.

First he wrote to Stirling, outlining his experience in mine explosions and asking for a blueprint of the Hillcrest workings. He next wrote to Campbell on July 4 to offer his expertise if the

Alberta government required it, and commented that the reports in the *Vancouver News-Advertiser* intimated that the inquiry would end as most such inquiries did, without determination of the cause. He wrote his third letter on July 11 to Minister of Public Works Charles Mitchell at the conclusion of evidence in the Hillcrest inquiry. It was the longest of his letters and outlined matters that he believed should have been discussed at the inquiry. He criticized the lack of information about the safety lamps recovered, the storage of explosives, the danger from roof falls, and the threat from electrical equipment in the presence of firedamp. He commented that no mention was made of explosives recovered after the explosion and observed that it was odd that some men's boots were burned, because as a rule, men are not burned lower down than the waist, and very seldom down to the knee. Presumably this comment arose from the fact that methane, which is lighter than air, collects near the roof; therefore, burns caused by the methane explosion would tend to be higher on the bodies of miners. However, if coal dust were disturbed on the floor by the methane blast, then the coal dust explosions would tend to burn the miners' feet and legs. Perhaps Ashworth's most important comment was that in such inquiries the commissioner ought to make use of one or more experts who are entirely independent of the inspector of mines, the owners, and the miners. He may have been correct in this comment, but Ashworth's other comments were based on incomplete information. Stirling duly replied to Ashworth's question about lamps with the information that they were Wolf bonneted safety lamps with two gauzes.

Ashworth was not satisfied with these answers, however, and wrote a second letter to the minister dated July 24, in which he asked for a copy of the mine plan showing the positions in which the bodies were found, the places where they were at work before the explosion, and a detailed list of the injuries and causes of death. He stated quite firmly that "the whole of the important evidence which might have been given to your Commissioner was not produced, not from any fault of the Commissioner but probably due to the non-appreciation of possibilities by whoever had charge of the government case." It was a criticism of Campbell's management "of the government's case." Ashworth concluded that "the importance of ascertaining 'the cause' is not the prosecution of the management, but to guard against the possibility of similar occurrences." Ashworth's search for answers was not an attempt to point fingers, but to stop such a disaster from happening again.

In another letter to Stirling, this one from August 14, Ashworth makes another valid comment. "It would appear to me that as the Commissioner was holding a Court in which Yourself and your staff, the mine owners, officials and miners were all treated as defendants, and as he was not assisted by a mine assessor — he could not be expected to make a practical enquiry such as he might have made had he had yourself as his assessor — it is the little points of daily mine practice which are of greatest importance and yet the purely legal mind does [not] recognise these."

It appears that Ashworth did not correspond directly with Brown, Hillcrest's general manager. Stirling, however, wrote to Brown for answers to some of Ashworth's questions. Brown replied that he could not understand what Ashworth wanted with the names and locations of bodies found with their feet blown off, although he supplied them to Stirling anyway. Brown stated that the best way to treat the subject was with silence. Campbell also did not reply to Ashworth's first letter about the disaster. From Stirling's polite but distant responses, one may conclude that he, too, had no interest in Ashworth's opinions.

Ashworth was clearly a man with a passionate interest in coal mine explosions, as well as some considerable knowledge about them. His letters display a deep dissatisfaction with the way the commission of inquiry and coroner's inquest were handled, the lack of expertise brought to bear on the matter, and the lack of key information. On the other hand, most of the officials involved in the aftermath of the Hillcrest disaster had some degree of knowledge of Ashworth himself, either from his time with the Crow's Nest Pass Coal Company, or from his involvement in the coroner's inquest into the Bellevue disaster in 1910. There, his theory had been accepted by the jury as the actual cause, due in part to his persuasive manner. But the theory was generally considered poppycock, or at least incorrect, and as a result he had become *persona non grata* at Hillcrest.

CHAPTER 13
A GREAT WASTE OF TIME

The coroner's inquest under F.M. Pinkney had been adjourned until the commission of inquiry had concluded its hearing and all of the recoverable bodies had been removed from the mine and viewed by the jury. The June 23 letter to Pinkney from Deputy Attorney General John Hunt unequivocally stated "that every facility should be given at the inquest to locate the cause, if possible." But since the inquiry had been held and managed so thoroughly by legal counsel for the various interests, what the coroner's inquest might do to upset the fine balance of competing interests had become a problem. In particular, the counsel for the Alberta attorney general, William Campbell, who had attended the commission hearings, wanted a brief inquest. In a letter to Hunt, dated July 15, Campbell said he would "endeavour to do all I can to shorten the proceedings although I am afraid the Solicitor for the Miners' Union and the Solicitor for the Italian Consulate are going to try to make the matter as full as possible."

In his letter, Campbell presupposed the outcome of Carpenter's report and wrote that "the evidence in the Commission has undoubtedly shown there was neither criminal negligence [nor] foul play." He added that in his opinion the coroner's jury had no role to play except to ascertain the cause of death and the identities of the men killed.

Hunt then wrote to Pinkney to tell him that in view of the complete and exhaustive inquiry into the cause of the disaster, "[I]t would be a great waste of time and money to go over the ground again at the inquest." He strongly advised Pinkney to confine his investigation to the determination of the causes of death and the identity of the men killed. In Hunt's words, "The expense has already been great and the sooner you can bring the inquest to a conclusion, after having ascertained the cause of death and identity of the parties, the better for all concerned."

Pinkney had now been placed under considerable pressure to heel, even if, having been present at the commission of inquiry, he had come to his own conclusions about whether the right questions had been asked of the right persons during the proceedings. Hunt wrote to Campbell to tell him of his letter to Pinkney, and expressed his hope that Pinkney would have the strength of mind to "hold down the parties who would seek to prolong the investigation."

The inquest began and ended in controversy. Pinkney opened it on July 21 at 10 AM in Hillcrest's Masonic hall. Court reporter T.A. Powell took the testimony of fourteen witnesses over two days. Present were legal counsel for the Province of Alberta, Campbell, Stirling, Aspinall, and the counsel for the UMWA District 18, Palmer. For some unexplained reason, MacLeod, legal counsel for Hillcrest Collieries, was not present, perhaps as a signal that the proceedings were considered unimportant in the context of the inquiry, which officials now claimed had been thorough. The solicitor for the Italian Consulate, Kappelle, was also absent. Almost from the beginning, Campbell, who had displayed a barely concealed contempt for Judge Carpenter at the commission of inquiry, also did not hide his contempt for the inquest and its jury.

William Adlam testified first. He said that he found the ventilation good, the mine just as usual, and nothing that morning raised his apprehension on the morning of the explosion. He also said that he did not have time during his shift to examine the whole of the mine. He had asked the mine manager, Quigley, to increase the number of fire bosses so that the entire mine could be examined regularly and Quigley had said he would see about it. The mine sat idle for two days with no one inside, on the 17th and 18th. In his examination previous to the two days the mine was not working Adlam said he examined only the entries and raises, not the faces, which meant that gas could have accumulated in the rooms at the face. He also said he did not know if, on the 19th, the brattice men had gone in to remove gas before the other men had begun to work. This was an extremely important admission of ignorance.

Unlike his testimony at the commission of inquiry, Dr. Ross now testified that the majority of the men died from gas poisoning. In response to questioning by a juryman, Ross said he had heard some doctors say that one Pulmotor was in good working order, but that they would not use another because it was not working properly. Campbell objected to the direction of the questions. "As I understand," he said, "this inquest is to find when, how and by what means certain people came to their death, and really I do not know whether it is in the scope of the jury — that is the first I have heard of this Pulmotor business." The provincial government carried responsibility for the rescue equipment, and Campbell did not want any fingers pointed at the province.

Later that day, John Ironmonger was called and asked who his brattice man was. He responded that it was William Twohey, but that he had not spoken to Twohey before the explosion. Campbell asked, "You do not know what he did with regard to clearing out the gas by brattice?"

Ironmonger replied, "Of course he went to the places where the gas was reported." But this was an assumption, nothing more. If Ironmonger had not spoken to Twohey, he had no idea if his brattice man had even gone to remove the gas. This outrageous statement of supposition was ignored by Campbell, who instead asked Ironmonger if he noticed anything in the mine that morning that caused him apprehension about safety. Ironmonger replied that he did not.

Campbell then asked Ironmonger to describe his escape from the mine. The fire boss replied that he was deafened, his ears were blocked up, and he staggered. Everyone around him was thrown into confusion. He began to run with others towards the entry. Men, he said, were dropping out of chutes until quite a crowd of men were running together. Then they met the heavy smoke. They kept going, but were nearly overcome until the air started improving. Ironmonger said if they had to go another fifty feet through the heavy smoke, they would have all gone down.

As soon as he got out and rested a few minutes, Ironmonger said, he went back down slant no. 2 to the north entry, where he had been at the time of the blast. No. 2 north was still full of smoke and afterdamp. He heard some men groaning and went to find them, but was just about overcome by the smoke and gas, so he walked back up to the surface and reported where the men he had found were located. Less than an hour later he went down again and found a man still there, sucking in water, so Ironmonger pulled him out. This other man was Gus Franz, who later died.

Ironmonger was asked if he had examined the brattice men's lamps, and said he wasn't sure. This was a curious response, and a complete contradiction of his testimony at the commission of inquiry, where he had said that he did examine the lamps. If we give him the benefit of the doubt and assume that he realized that his earlier answer was incorrect, then his answer at the coroner's inquest also presents a problem: if the brattice men had gone into the mine half an hour before the miners, he would surely have remembered that he checked their lamps. It is more likely that he would have forgotten had they had gone in at the same time as all of the other workers.

Twohey was later called as a witness, and stated that the ventilation was running contrary to the Mines Act that morning, as it had been switched to blow air into the back airways, which then came out of the no. 2 slant. Both Campbell and Palmer ignored Twohey's comment, but Aspinall did not. He asked Twohey about his knowledge of the Act, and how the reversal of the fan violated it. A juryman then asked Twohey why, if he knew that it was a violation of the Mines Act, had he not reported it? Twohey replied that he had reported it innumerable times, and the pit boss, Thomas Taylor, told him that he knew he was violating the Act. Twohey added that he thought Taylor fancied that the change would give better results, and it did.

Twohey testified that he had found the fan stopped once without notice due to mechanical failure, with the result that there was no ventilation. He had gone to the telephone at the bottom of no. 1 south and rang up the engineer to tell him the fan had stopped. His call was answered by the brother of the manager, John Brown. The pump boss told Twohey there was a man down there, in the mine, to look after the fan. Brown obviously mistakenly assumed that Twohey was talking about the hand fan, and Twohey told him that the fan was not in the mine, but up on the surface. Brown told him to go to hell, and then hung up. This occurred two and a half months before the explosion. Twohey said that once the fan stopped, the gas quickly accumulated. Nobody called to tell them that they should get out of the mine, so Twohey himself had told the miners that the fan had stopped. They got out and went home.

Twohey was asked if the reversal of the fan after the explosion had helped men to get out. Contrary to previous testimony at the inquiry, he said it had not helped. Twohey's testimony, however, was far more important for what he was not asked. He was not asked directly whether a fire boss had given him instructions about clearing gas, if so, which one, whether he had cleared gas, how had he done so, and where. He was not asked if he was with the other brattice man, David Harris, when or if Sam Charleton had given them instructions to remove gas. He was not asked what time he had gone into the mine, or whether he had seen what time Harris had gone in. The failure to ask these questions was an extraordinary oversight by the officials present. Twohey did, however, say that Charleton had examined his lamp before he went into the mine on the morning of the accident.

The apparent lack of interest in questioning an important witness like Twohey at the inquest, which had been intimidated into moving quickly, was only an echo of the orchestrated lack of rigour in the allegedly thorough inquiry, which had not bothered to question the only surviving brattice man at all.

On July 22, Pinkney read the letter from Deputy Attorney General Hunt, which insisted that the inquest confine itself to the identity of the men killed, and their causes of death.

Campbell then gave a long address in which he argued for an immediate conclusion to the inquest, which he said was "not only superfluous but I think it is utterly — well I do not say uncalled for — but it is most deplorable that there should be such a waste of expense and waste of time and recriminations and matters which do not affect anybody and cannot affect anybody, because your worship will readily understand that the object of a coroner's inquest is not to punish anybody or to make any findings that have any weight in law but simply to make an investigation into which evidence comes forth and on which the authorities can act."

One of the jurymen reminded everyone present that shortly after the explosion, Pinkney received a letter from the attorney general's office saying that the inquiry should be of the fullest possible nature. Now the same office, in another letter to Pinkney, countermanded that request by stating that the commission had done its job, and all that was left was to find out by what means the victims met their death. The juryman was clearly angry. "I never heard of anyone giving suggestions to a coroner's jury," he said.

Campbell was adamant, too. He insisted that if the government thought an inquest was uncalled for, they could stop it from proceeding altogether. In his rambling, almost incoherent address to the jury, Campbell accused the mine inspection department of making the companies stand around, insisted that the jury could not find evidence as to what happened a month or two earlier in the mine, or of what had happened since. Campbell, apparently in a nearly incoherent rage, said, "You have simply got to find what was the cause of that particular explosion." He concluded, "I think it is my duty to say here that to go into matters which do not affect this accident, which do not affect the death of these men, or the cause of death, is an insult to the Commission. That is all I have to say about it."

In fact, it was not all he had to say about it. A juryman told Campbell, "It is not your purpose to curtail any of the evidence we might want to arrive at the purpose for which we are here." To that, Campbell responded that it was his duty to say that if an inquest is called to inquire into the death of certain men, and evidence to find out something else is introduced, it is his "duty to stop it, because they have no right to listen to it."

Pinkney, at that point, backed down. He suggested that there might be some misunderstanding about the role of the inquest, and that in light of the establishment of the commission, he would not have been surprised if the provincial government had called off the inquest. As a result of the commission having gone into the disaster thoroughly, he said, the duties of the jurymen had been lessened "somewhat." Pinkney agreed that only a few more witnesses should be called, and the inquest should then be wrapped up.

William Jolley was called. He was with the Frank rescue team that included Hallett, Burrows, and Leavitt. It took, he said, three quarters of an hour to get the rescue apparatus assembled.

Jolley testified that men went down to the rescue car five times to get parts, and all the while they were trying to get the apparatus together, men would come to them from the mine and say that they had seen men waving their lights inside, looking for help.

One of the jurymen asked if he had seen the Pulmotor in use, and if it had been working properly. Jolley responded that he could not get to it because the police were keeping everyone back.

Palmer then asked if he had seen men waving their lamps for help in the mine firsthand. Jolley testified that he had not. Those who had seen lamps, he said, had not seen the lamps of miners inside who wanted help, but rather, had seen the lamps of the fire boss, Adlam, Dodd, and another man at the top of the slant who had been stopped by heavy smoke from entering to find survivors. He added that the fact that the apparatus was out of order had no effect on saving or not saving lives.

Under Aspinall's questioning it became apparent that although Jolley had been trained in the use of the rescue apparatus, he seemed to have forgotten it. In fact, Jolley had not received a certificate of proficiency with the apparatus, and had taken only eight days of training instead of the required ten.

The Frank rescue team was late getting into the mine. Once inside, with the apparatus and the oxygen turned on, one of their team members lost consciousness. His bag had filled up too quickly, and he and some of the rescue team had begun choking. Fred Elliot, in a panic, had left the team and fled up the slant, which meant the remaining men were unable to lift the unconscious Leavitt. It had nearly been a disaster.

Pinkney asked if, even with all the apparatus in working order, any additional lives would have been saved. Jolley answered, "No."

Campbell, who had been silent, suddenly spoke up. "Yes, there is a witness has sworn that no lives could have been saved anyway, and now they are enquiring further; it is too ridiculous."

Willam Goodwin was sworn in, and testified that the Frank rescue team was not trained well enough to understand their limitations when using the apparatus. Men had not been trained on any inclines, just on the level, and so did not know what they could do with it, since more oxygen is consumed on grades. His opinion was that if the rescue team had gone right into the mine, there was a possibility that some of the men might have been saved.

The man in charge of the rescue car, Henry James, was called to the stand. James testified that once he arrived with the rescue car from Blairmore, he had to wait for wagons to take the apparatus up to the mine. He said that he sent five complete sets of apparatuses up to the mine. Another five sets that were initially left in the rescue car just needed their reducing valves to be attached to the bags, and the gauges attached to the tubes. These were retrieved from the rescue car later.

James was asked if anyone came down to get gauges for the apparatus, but he could not remember. He said he left the car when it was convenient and came back three or four times for different things. James said he sent two Pulmotors up to the mine, but found only one of them later in the lamp room. The other had gone missing. He said the one he discovered did not appear to him to have been used.

When asked how long it would take to attach the valve, James replied it might take ten seconds. James said that during the rescue work two men who had both had training some months before asked to go into the mine with the apparatus. He showed one of them how to operate the valves four times, and he still could not understand, so James refused to let the man use it. He refused the other man as well. James said he trained men for ten days, and would not give a certificate unless the man demonstrated a proficiency with the apparatus.

John Stirling was sworn in. He was asked if provisions were made for searching men for matches and tobacco. Stirling said it was not compulsory, but that a manager could make a search or appoint someone. There was no provision in the act to compel a company to keep a record of such searches. Stirling said he thought it would be a good idea to make such searches compulsory.

Aspinall was called, and questioned by Campbell. He testified that the dust in the mine was worse now than it had been about a year ago when he had last inspected it, that the ventilation at the time of the blast was inefficient, and that if he had seen gas being removed over working men in the quantities reported, he would have had the men leave the mine until the gas was expelled.

Campbell asked Aspinall if the change of the direction of air flow at the fan was a good idea. Aspinall replied that it was good in some ways, but not in others. The only problem with it was that too many men were in the area affected by that air flow.

Aspinall changed his testimony considerably from what he had said at the commission of inquiry. He had testified then that he considered the change to be seriously flawed and that he would have shut the mine down or prosecuted the company for the reversal of the fan's air flow. He had also accused the fire bosses of lying. It was a shocking change that illustrated the pressure that had come to bear upon him, or a change in his knowledge. Either way, he did not look good.

J.G.S. Hudson was sworn in, and was asked by Campbell about training in use of the rescue apparatus. Hudson responded that if a man gets excited or walks up a steep incline, he uses a great deal more oxygen than he would on a level surface. He said it takes a man with very cool nerves in situations of this sort to keep himself under control for his own safety. Hudson pointed out that men can be trained, but until they get into an actual rescue situation, it's difficult to know how they will behave. Training, he said, doesn't alone make a man experienced; it also takes aptitude.

Hudson said that in his experience after a disaster, there is never a difficulty in getting volunteers from other mines. The difficulty sometimes is in controlling the men who are anxious and willing to go, and choosing who is best suited for the work.

At 4:05 PM the jury retired to render a verdict. At 4:45 they returned, and the foreman read the verdict aloud:

> These men came to their death in the Hillcrest Mine on the 19th day of June 1914 as a result of an explosion of gas and coal dust.
>
> The jury desire to add that they do not think the regulations of the Coal Mines Act have been strictly adhered to.
>
> We would recommend that the Government enforce an inspection at intervals of not less than once a month for matches and pipes of all men employed in the mine.

The jury further recommended that each company keep on hand, as near the mine mouth as possible, a sufficient number of safety apparatuses in case of accident.

The verdict was signed, "Frederick M. Pinkney, Coroner."

CHAPTER 14
UNFINISHED BUSINESS

The loose ends were almost all about money and payment for services after the disaster, and the invoices began to come in. On July 13, the manager of Jasper Park Collieries, Andrew Scott, who had given testimony at the commission of inquiry, mailed an invoice for his time and expenses to Crown Prosecutor Campbell at MacLeod. It included $42 for the train fare, $40 for ten days' personal expenses, including hotel and meals, and $150 for ten days' fees as an expert witness, for a total of $232. Campbell was outraged, and wrote to Stirling to inform him of the "absolutely absurd" amount.

The invoice for the hall rental from July 2–11 was passed on by Corporal Mead of Bellevue to Inspector Junget. It totalled $70.70.

The cost of the coroner's inquest, including fees paid to the coroner, jurymen, witnesses, and two days' rental of the Masonic Hall, totalled $413.70. It was passed on to the attorney general's department for payment by the treasury.

On July 29, the commissioner of the RNWMP requested that the exemplary conduct of Corporal Mead, Corporal Grant, and Constable Hancock during the disaster be mentioned to the prime minister, and that the comptroller of the RNWMP sanction a grant of $50 each in recognition of their work. RNWMP relations with miners and their union were never amicable, but the work of Mead, Grant, and Hancock had earned respect among the miners, citizens, and officers of the union who had actually praised the Mounties for their attendance to the bodies. Junget quotes these union officers, whom he describes as "the rankest socialists and the enemies of any police or military forces," as saying, "We have no use for the police, but we cannot help respecting its members when we see them working under such trying conditions." These conditions included handling the mutilated and battered bodies, a "gruesome" job. Many of the bodies, Junget pointed out, had been mutilated, with both legs and heads blown off, and the men had been severely taxed by the experience. Along with volunteer miners, Mead and had stripped the bodies, washed them, and wrapped them in white cotton cloth. The request for the grants, an amount equal to about three months' pay, was approved.

The Union Hotel had supplied hundreds of meals for government officials and Alberta rescue teams during the days following the explosion, and the government now had to pay for them. The new proprietor of the hotel, Charles Fuchs, mailed an invoice to John Stirling for $140.50 (or 281 meals, at $0.50 each), a cost that Stirling considered outrageous. In his reply to Fuchs, Stirling called the invoice exorbitant and suggested that because there were so many meals, only a lower rate would be acceptable. In the end, Fuchs lowered his cost to $0.40 per meal, which satisfied Stirling.

By July 31, within six weeks of the disaster, 140 men had arrived to take the jobs left vacant by the fatalities among the Hillcrest miners, and a new mine manager, R.T. Stewart, was appointed to replace Quigley. Sam Richards replaced Thomas Taylor as overman, or pit boss. Robert Hayes replaced Sam Charleton as fire boss, or examiner. Among the new men were many of the miners who had lost their jobs in the closure of the CPR mine at Hosmer.

One man stands out among those who rushed to Hillcrest to assist in the rescue and recovery. On August 11, a miner named Alexander Muir wrote a polite letter to Stirling, reminding him of a conversation they had at Hillcrest. Muir was employed at the no. 6 mine at Hardieville near Lethbridge, and did not hear about the explosion until after the special rescue train had already left. Muir had trained with the Lethbridge team, and knew that he could contribute to the rescue and recovery efforts. In his desire to help, Muir paid his own fare to Hillcrest and hopped a train before he had a chance to change into his work clothes. The first man he met on site at Hillcrest had been general manager Brown, who asked him if he was trained in the use of the rescue apparatus. Muir explained that he was, but first wanted to find a pair of overalls because he was wearing his good clothes. Brown had told him not to bother with that, since the others on the team were waiting, and if his clothes were damaged, they would be replaced.

Muir went straight down into the mine and, as might have been predicted, badly damaged his good clothes. Back on site, Muir reminded Brown of his promise, who told Muir to make sure that Stirling was informed about the matter. Muir did so, and Stirling made a note of it.

Muir had written to Brown earlier in the summer about the replacement of his clothes and compensation for his fare, but had received what he considered an unsatisfactory reply, so wrote to Stirling instead. In Muir's words, "I think it will be rather hard times on me if I do not get my clothes replaced after losing them in such circumstances." He asked Stirling if he remembered their conversation or still had his note, and if so, to pass on a letter to that effect so that he could take it up with Brown.

In his reply to Muir, Stirling said that he remembered the conversation, but was not aware that Brown had made any promise about the fare and the clothing. Stirling also sent a copy of the letter to Brown, but any correspondence between Brown and Muir is lost, so it is impossible to know if Muir ever received the compensation he sought. Muir's letter and Stirling's reply illustrate how an individual of no influence could be forgotten in the big picture. Muir, a man of very modest means, had wanted to help, had the skills to help, and was gladly used as a resource by the authorities. He did not ask to be paid for his time. That he had to virtually beg for compensation for his ruined clothes and fare seems a niggardly and small-minded way to treat him.

For some members of the public, the name of a victim published in newspaper accounts could give rise to fears for a loved one with a similar name. A woman in Winnipeg was such a case. She feared that a man named David Gwyllan Thomas, who was supposed to be working in the mines at the time of the Hillcrest disaster, might have been one of the victims. Her fears, however, were unfounded. Hillcrest Collieries wrote a letter back, informing her "that only one man of the name of Thomas was killed, viz., David O. Thomas, an American, who has a number of relatives in Montana."

As a result of the disaster, and the fear that shot firing might have been its cause, the management of Hillcrest Collieries completely stopped shot firing in the mine after resuming operations. The commission of inquiry had all but eliminated shot firing as the cause of the disaster, but the level of fear had risen and no one, management or government, wanted to be held responsible for a repetition of the disaster. It was obvious that coal dust had contributed greatly to the intensity of the explosion, so the one possible source of coal dust ignition, apart from a methane blast, had to be eliminated. This did not sit well with the miners whose pleas to allow shot firing were ignored. Work became much more difficult when they could not shot fire. By mid-August, the Hillcrest miners walked out on strike because of it.

One of the more important documents related to the disaster was the fire bosses' report book. It had been in the possession of Judge Carpenter among the documents presented as exhibits at the inquiry. Hillcrest Collieries wanted it back after the report was released, but it had gone missing. On May 5, 1915, Carpenter wrote to Stirling that he was unable to find the book, which contained the record of shots fired and inspection reports for the Hillcrest mine. "It is possible," he writes, "that it has got in with the other books that were put in as exhibits." Carpenter continued that he was certain that it was turned over to the district court clerk, or had been "unfortunately lost." The judge apologized for its disappearance. All that remains of the inspection reports are handwritten copies, none of them notarized. The copied reports are accurate, the veracity of their wording authenticated when they were read aloud at the commission of inquiry, but the loss of the originals remains another of the mysteries of the disaster.

Among the many questions that remain, and which will perhaps never be answered, is the issue of the number of bodies still buried in the mine. Officially, one body could not be recovered: that of thirty-eight-year-old Sidney Bainbridge. But William Hutchison tells a different story in his unpublished memoir.

Hutchison is specific. "All the bodies were got out except two," he wrote, "a chap called Bainbridge, and his partner." However, the list of bodies recovered lists Bainbridge's mining partner, Fred Bingham, among them. Hutchison writes that Bainbridge's two brothers-in-law, who were miners at Coleman, asked Brown for permission to go into the mine and attempt to retrieve Bainbridge's body. Brown approved the request, but asked Hutchison, who knew exactly where the bodies were, to go with them.

Hutchison says the Coleman men, both experienced miners, did a good job of shoring up the heavy rock fall that had buried Bainbridge and his partner, and that the chances of recovering the bodies within a few shifts looked good. However, that night another shift came on, and when Hutchison and the two men returned in the morning they found that their work had been compromised by the other miners. They had just started to get the shoring back into place when the district inspector came in and told them the night shift had reported bumps, and said the place was unsafe. The chief inspector then ordered their work stopped altogether rather than risk the lives of anyone else. Hutchison and his two companions were left disappointed and angry. Hutchison dealt in fact, not fabrications, and his veracity, even in old age, made it difficult to question his claim that two men, not one, were still buried in the mine.

On June 26, the company released its final official list of figures. It claimed that three bodies remained in the mine: Oakley, Bainbridge, and an unknown. On July 6, a body was recovered and identified as Oakley. At 12:30 PM on July 14, the body of William G. Miller was recovered from the mine. It had been assumed that Miller had already been buried. Up until then it was believed that the missing man was Bolinski, but it became clear now that Bolinski had been buried as "unknown." Three days after the recovery of Miller's body, Stirling wrote a letter to the editor of the *Labour Gazette* in Ottawa, who had asked for details about the disaster. Among the details Stirling passed on was the comment that "[t]here are supposed to be still two bodies in the mine." Possibly, only three days after the discovery of Miller's body, Stirling had not been informed about it, but the figure more likely bolsters Hutchison's claim that Bainbridge and another man remained in the mine.

Hutchison believed that two men, not just one, were buried in the cave at the face of room 59. Is it possible that the man buried in a grave as Bingham was actually someone else? Presumably, company officials knew the total number of men in the mine, and they knew who escaped alive, and who was killed. Obviously, they believed that they had buried 188 men, but in some cases buried fragments of bodies and made assumptions in the absence of positive identification. Certainly the accurate identification of bodies had been a serious problem for teams at the wash house. Some bodies had been blown to pieces, or crushed beyond recognition. Others had clothing blown or burned off, or had lost their identifying checks. Still others were found badly decomposed a considerable time after the explosion. Hutchison does not name the man he alleges to be buried with Bainbridge, and refers him only as Bainbridge's "partner." In his memoirs, Hutchison underlined the final number of dead as 189, so we may take it that he was aware that the suspicion might arise that the mystery man allegedly buried with Bainbridge would bring the total number of dead to 190, and that he wanted to dismiss such a suggestion. The confusion must then arise in the counting, identification, and assembly of parts of bodies, or in Hutchison's memory. One is forced to wonder why Hutchison did not say *why* he believed that two men were still buried in the mine. While we might conclude that the likely identity of the alleged second man is Bingham, a company document and newspaper reports indicate that Bingham's body was returned to Nova Scotia for burial. The truth in the matter remains elusive, and we are left to ponder another of the lingering mysteries that surround the Hillcrest Disaster. The most important mystery, of course, was the one that the province had employed judge Carpenter to solve.

CHAPTER 15
CONCLUSIONS

Judge Carpenter delayed his report on the Hillcrest disaster until the fall, while he awaited the results of tests on coal dust. The dust samples had been sent to the U.S. Department of the Interior's Bureau of Mines Experiment Station at Denver, Colorado, because there was no facility for such testing in Canada. On October 20, Carpenter mailed his report of the commission of inquiry from district court in Calgary to Minister of Public Works Charles Stewart in Edmonton. The results of the tests proved what had already been known anecdotally: the Hillcrest coal dust was highly explosive.

Carpenter's fifteen-page published report gives a clear and concise outline of the evidence presented at the inquiry, as well as his conclusion as to the cause of the disaster. He observed that all explosions, except those caused by a blown-out shot, originate with the ignition of gas. A blown-out shot had been ruled out by virtually all of the witnesses, so the question became: what ignited the methane? A defective safety lamp was possible, but unlikely. As for the possibility that someone lit a match, Carpenter said such a scenario could not be eliminated as the cause. He pointed out that although mine management had the right to search men for matches, pipes, or tobacco, there was no evidence that such a search had been made. Although Carpenter did not state so in his report, no one aside from the general manager, John Brown, was questioned about whether any searches had been conducted — and the lawyer for the company, MacLeod, had supposedly established that Quigley, not Brown, was responsible for such matters. If Brown's job was not to supervise mine operations, including safety procedures, then what was it? It is strange indeed that Palmer neglected to ask such a pertinent question. No fire boss was questioned about searches for matches or smoking materials, yet three fire bosses and a fourth, former one would have known the answer. It was, one must conclude, a serious failure on the part of virtually everyone involved in the commission of inquiry, including the RNWMP, who seem to have stepped into the shadows and watched events unfold from the perimeter.

The issue of miners smoking in the mine becomes even more curious when the conclusions of the coroner's inquest are examined. The coroner's jury recommended "that the Government enforce an inspection at intervals of not less than once a month for matches and pipes of all men employed in the mine." No testimony indicated that smoking was an issue, yet the jury made a recommendation that seemed to come from nowhere. Why? What closed-door discussions might have suggested such a recommendation? Was something found on the bodies of the dead miners that might have indicated miners were smoking? The answer might be as simple as the fact that Stirling made such a recommendation at the inquest. But it was as odd a recommendation as suggesting that no miner be allowed to walk into the mine with dynamite in his hip pocket. Such a recommendation would not be made without a reason to suggest that it was an issue. No reason here was ever given. The sentence puzzled others at the time, too.

In his correspondence with Stirling, James Ashworth commented in a letter dated August 14, "The recommendation of the Jury with respect to matches and pipes must have had some basis, and therefore I anticipate that both these articles were found on some of the bodies as at Bellevue — might I trouble you to let me know if this were the case? And on which bodies matches were found?" It was a reasonable question.

In his reply, Stirling said, "[A]s far as I have been able to ascertain and as far as the evidence shows, there were neither matches nor pipes found on any of the bodies. It is therefore difficult to account for this rider which was added by the jury." Difficult indeed. The general unwillingness to pursue the matter at the inquiry, and now this jury recommendation, were certainly cause for concern.

The mine plans on which Stirling had drawn the direction of force in the explosion, as far as he was able to determine it, and which other experts agreed was accurate, are remarkably complete, and leave little doubt as to the origin of the explosion: the second crosscut between rooms 41 and 42, off the no. 2 south level. From there, the lines of force go both ways. There is no rock fall there that might have pointed to sparks as the cause of the explosion. A high-pressure burst of gas that could conceivably be ignited by a safety lamp tended to happen at the face, not at a crosscut, but not a single witness had so much as suggested a burst of gas as the cause. Experienced fire bosses Ironmonger and Briscoe both testified that they had neither seen nor heard of blow-outs of gas in the Hillcrest mine. Furthermore, a lamp would likely not be placed at the crosscut; it would be at the face. The only reason for a source of ignition to be at the crosscut between rooms 41 and 42 would be if a man had gone there to smoke, away from the face where they were working, and where gas would be most likely to accumulate.

Another piece of the puzzle surfaced many years later in an interview conducted by CBC Radio with Hillcrest storekeeper George Cruickshank. A sort of town confidant among miners and their families, he knew literally everyone in town, and as the Worshipful Master of the Sentinel Masonic Lodge, which was located above his store, Cruickshank became a clearinghouse for information of all sorts. He was a community leader, generously gave credit at his store to sometimes hard-pressed miners, and was deeply trusted by miners and their families. In the interview, Cruickshank is asked about the cause of the Hillcrest explosion. He replies, "We all have our ideas how it happened, and I think my idea is correct, but I wouldn't like to mention it." Why would he not reveal his thoughts on the cause? One concludes that he had sworn not to tell something he had been told in confidence, or perhaps that someone would have been incriminated. Otherwise, why not reveal it? Cruickshank's reluctance to speak pointed to individual responsibility. There is only one cause that could be attributed to a single individual: smoking. But no one wanted to discuss this, because it would have indicted an individual, the union, and the company. It was also, given the circumstances of mine explosions in general and the crude forensic techniques of the day, virtually impossible to prove.

Ventilation of the mine, another responsibility of the company, was also a serious concern. Evidence showed that the raise in no. 2 south was full of gas, and that the fan boys who turned the two hand-operated fans in the raises entered the mine at the same time as the miners, rather than half an hour earlier with the brattice men, as would have been prudent. The later entry meant that the fan boys would blow gas from the raise *over the men working at the face,* a practice that could only be described as poor. On the other hand, as Carpenter pointed out, the working men testified, virtually unanimously, that the ventilation was good in the places where they worked. It was rare for miners to support their company's claim that it was living up to its management responsibilities.

Carpenter wrote in his official report that he was puzzled by the key testimony of one fire boss:

> I confess that the evidence of Adlam somewhat bewildered me in regard to the quantities of gas referred to in his report. My impression from his evidence given in the first instance was that there were comparatively only small quantities of gas in the places indicated in his report. Upon his being recalled his evidence gave me the impression that the quantities of gas were much greater than his evidence led me to believe in the first instance and I am somewhat at a loss to reconcile his different statements in this regard. Adopting a later statement, it is evident that the raises in Number 1 North Level, and in Number 2 South Level, were, as he says, full of gas. With these raises full of gas it does seem that it would have been advisable that these places should have been cleared before the miners entered the mine. Such a course at least would have avoided an element of danger that had to exist if the raises were being cleared after the miners had gone to work.[1]

Carpenter points out nonetheless that Adlam said he did not consider there to be an unusual amount of gas in the mine that morning. The key word here, of course, is "unusual," which implies there was always a good deal of gas in the mine. Even Hudson, the experienced mining man, said that there was nothing indicating an *undue* amount of gas at that time. The obvious problem with these adjectives was that their vagueness left them open to loose interpretation, and therefore made them excellent tools for evasion. Hudson's supposed experience in mine explosions was also an issue that may indicate why he spoke in vague terms. It was called into serious question by the self-proclaimed expert James Ashworth, who knew Hudson and commented in a letter to the Alberta minister of public works that Hudson in fact had very little experience in finding out the cause of such disasters.

A critical review of the two inquiries and their conclusions reveals serious deficiencies. These include evasive answers by witnesses, a consequent failure by the lawyers to pursue unequivocal answers, failure of the commission to call certain relevant witnesses (and, we may surmise, failure of these witnesses to step forward), and failure by the lawyers to ask certain pertinent questions of particular witnesses. All of which contributed to an evasive environment with only the semblance of earnest inquiry.

In particular, the lawyers failed to pursue clear, unambiguous answers from the fire bosses about the brattice men, whose work was to remove dangerous gas from the mine, and who were under their direct charge. The lawyers also failed to ask about previous searches for matches and tobacco.

If Alex May had been at the no. 1 fan to inspect or repair it, as reported, why was he not called to testify as to its state of repair? Why was the man who operated the no. 1 hoist in the hoist house, which had its roof blown off, not called to testify, whereas the hoist engineer at the undamaged no. 2 hoist was? Even the name of the no. 1 hoist operator was not mentioned at either the inquiry or the inquest.

Carpenter's report concluded "that the disaster was caused by an explosion of gas, the origin and seat of which is unascertainable, this explosion being augmented by the ignition of dust throughout the mine."

The problem with the inquiry was that it was controlled by lawyers at every turn: Campbell for the province, Palmer for the UMWA, and MacLeod for the company. These men decided what witnesses should be called. Their duties and responsibilities, however, were to their clients, and they did their jobs well. Unfortunately, in the avoidance of assigning any responsibility for the explosion, they also sacrificed finding a cause. There was no advocate for truth, in the abstract, larger sense of the word. Among Judge Carpenter's responsibilities was to enforce the expression of truth by those sworn to testify, but if witnesses with important information were not called, Carpenter was helpless. If the right questions were not asked, he was helpless. He could not know who had valuable evidence that might have a bearing on the cause of the disaster. If such individuals would not step forward on their own initiative and were not called by lawyers, or, if they were called, were not questioned rigorously by lawyers, then larger truths could not be revealed.

Responsibility, however, is tied intimately with cause. If a miner struck a match to light a pipe or cigarette, for example, a layperson might deem the lit match to be the cause, and the man who lit the match responsible for the ignition of firedamp. On the other hand, in a legal inquiry, not only would the miner be responsible, but also the brattice man who failed to properly test for and remove the gas, the company that employed him, and perhaps the fire boss who failed to ensure that the brattice man had done his work properly. This logic could, of course, be carried even further. If another miner saw the man light the match but did or said nothing to the fire boss, was he not also partly responsible? Did not the union carry some responsibility for its members and their safety? Did the union urge searches of its members? No, it did not. Nor had the union pit committee found anything wrong with the operation of the mine. Surely they too shared some responsibility. To untangle the intricate web of responsibility was not the mandate of either the inquiry under Judge Carpenter, nor the inquest under Pinkney. No one person or group was responsible for the obfuscation and lies, but rather the pragmatic approach to answering what was asked and said based on the projected consequences of a public airing of said truth.

Ultimately, everyone was responsible, and no one was responsible for the Hillcrest disaster. The dead Hillcrest miners and their families were victims not only of ignorance among the ranks of fellow miners and company men, but of human fallibility, of the hostility between miners and management generated by Socialist and Marxist union politics, of Canadian immigration policy which allowed so many unskilled workers into the country and eroded the value of labour, of the abundance of coal in the province which meant anyone could dig a hole almost anywhere and find coal, which decreased the value of coal itself. In the period between 1895 and 1962 in Alberta, more than 200 mines operated at Edmonton, more than a hundred at Lethbridge, about 135 at Taber, and more than 145 at Castor. Hundreds of others operated at Drumheller, Wetaskiwin, Camrose, Ardley, Carbon, Brooks, Champion, and Milk River.[2] Perhaps the only surprise is that disasters were not more common than they were.

One may ask whether the CPR's uncharacteristically speedy closure of the Hosmer mine immediately after the Hillcrest disaster was a political move, one designed to remind everyone at Hillcrest that their coal was needed, and that Hillcrest should be kept open, in part to provide jobs for the many newly unemployed Hosmer miners. To make matters worse, 1914 was a year of recession. The Hosmer closure therefore presented a tidy double problem for the factions, as the CPR knew it would: what to do about Hillcrest's widows and orphans, and what to do with the former Hosmer employees, most of them young men with families. Hillcrest wanted skilled employees to replace the dead miners. Hosmer miners wanted jobs. The CPR and Hillcrest Collieries wanted the mine in operation, the sooner the better. The UMWA wanted work for its members, assistance for the widows and their children, and to move on. The provincial government wanted order, peace, employment, a safe mine, and compensation to widows and their children. It was a rare time indeed when all the factions agreed, or at least had overlapping goals. Although there appears to be no evidence that the factions discussed these issues together, they were certainly discussed in closed-door meetings of the UMWA, the CPR, the provincial government, and Hillcrest Collieries. In order to arrive at an understanding, the issues need only have been discussed among the factions in a few private, casual conversations between individuals, unofficial of course, reinforced by the newspaper articles that appeared, particularly in the *Lethbridge Herald,* that touched on the relevant issues.

Each of the factions was acutely aware of the consequences of a bankrupt company. Widows and orphans would not see compensation, and become a burden on the union and the provincial government. The contract miners and company men would lose their jobs. The best coal in the Crowsnest Pass, coal that the CPR desperately needed for its locomotives, would stay in the ground until a new company could be cobbled together out of the ruins of a bankrupt Hillcrest Collieries. The shareholders in Hillcrest Collieries would suffer huge financial losses. On top of that, after fingers were pointed, some individuals with families would have their lives ruined, and their ability to earn income shattered. In the face of all this, responsibility for the disaster assumed a kind of irrelevance, and in truth an enemy to virtually everyone involved. Certainly the pragmatism embraced by the principals contained an irony that some must have noticed: the three factions, the union, the company, and the provincial government, and a fourth grey eminence, the CPR, which in the past were so often at odds with each other, now all had to agree, whether verbally or by unspoken understanding, that Hillcrest Collieries could not be allowed to close.

The supposed lack of knowledge about the explosive character of the coal dust, and the evasions of questioning about searches for matches and smoking materials implied that both management and union never seriously considered that the Hillcrest mine could explode. The union did not press for searches of its members, just as, years earlier, it did not press for the use of safety lamps. This was because its members did not want them, and union leadership was not inclined to push against the will of their own membership, a demonstration of the fact that unfettered democracy, without constitutional limitations or principles, is just mob rule. Their denial of awareness that the dust was explosive was part of the pretense that had they known how explosive it was, they would never have allowed any dust to accumulate. Neither the company nor the union made any request to have the dust scientifically tested before the explosion. Why? Because they knew it was explosive. Any experienced miner would know this. Hillcrest coal was high-grade bituminous coal, and its explosive nature was as certain as the law of gravity. To test its explosive character would be as absurd as Albert Einstein calling a news conference to test whether a dropped object would actually fall to the ground. Coal dust from far inferior coal, found on the ground in Edmonton's Mill Creek Ravine, near the former Twin Cities mine, for example, will explode when tossed into the air above a flame.[3] In fact, any mixture of air and a finely powdered inflammable substance — sawdust, grain dust, flour — is demonstrably explosive.

An editorial in the *Coleman Bulletin* of June 25, 1914, offered some timely observations with respect to the disaster. The editorial commented on Alberta's huge coal deposits and remarked on the tremendous loss of life in the Hillcrest disaster. The loss to families, it observed, is irreparable. It then commented that "surely the same science that is able to spy out these coal deposit … is also able to devise ways and means of coping with the many dangers incidental to the mining of the coal?" The next comment provides another clue as to the reason for solidarity among the four factions in the inquiry: "The government appoints many inspectors to guard against just such occurrences as the Hillcrest horror, and from the fact that these inspectors are appointed we take it that there are methods by which gas or dust can be detected in any mine before it has had time to cause any loss of life. If not, of what use are inspectors?" Indeed. The editorial casts responsibility upon provincial mine inspectors. It would be, however, impossible to apportion all of the blame to any single faction. As soon as a finger was pointed in one direction, another finger would be pointed in another. The battle of blame would leave no one unbloodied.

. To suggest that some brattice men and some fire bosses took a somewhat casual approach to their responsibilities, and failed to properly remove gas from the mine, was to suggest that management was incompetent, that miners who claimed the ventilation was good were either liars or fools, and that provincial mine inspectors had not done their jobs and were useless. To suggest that a miner had set off the explosion with a match was to suggest that miners were not all victims of ruthless mine bosses. The truth likely lay tangled somewhere in the Gordian Knot of human behavior that involved politics, the struggle for control, human failings, fear, shrugged shoulders, equivocation, evasions, and fatalism.

The failure of both inquiries to uncover the truth of what happened is clear; the failure of will to pursue the truth is perhaps somewhat less obvious, hidden as it was under a veneer of rectitude. The number of Italian miners killed, thirty-five, was so great that it prompted the Italian government to hire a lawyer to represent the interests of Italian miners at the inquiry, yet not a single Italian testified at either the inquiry or the inquest. Not a single Ukrainian testified either, although twenty-six died. In fact, not a single Hillcrest contract miner or company man who did not speak English as their native language was asked or subpoenaed to testify. Even Campbell stated that he had no doubt that several miners whose mother tongue was not English got out of the no. 2 south level, and that the commission ought to hear some of their evidence. It never happened. It seems difficult to believe that men such George Frolick, a Hillcrest veteran, could . not have offered valuable testimony. It is also difficult to believe that language issues could be used as an excuse when the UMWA commonly published notices in various languages and had the resources of translators. The union had, in fact, supplied a translator at the Bellevue inquest for a man who spoke no English. But no one made the effort to speak to these men. Ethnic divisions were rigid. The bosses in the mines, almost without exception, spoke English as their native language, and anyone who did not speak English well made it a habit to keep his profile low and his mouth shut, except during union meetings. None of these men would step forward to volunteer testimony that might place his job and livelihood at risk. To risk one's family's well-being was not an option. It was also in the interests of these men to keep the mine in operation, and not to upset what appeared to them to be orchestrated testimony.

CHAPTER 16
ECHOES

Judge Carpenter offered a number of recommendations based on his interpretation of the evidence submitted, among them that any mine ventilation fans, such as the no. 2 fan, should either be under constant supervision, or have an automatic indicator attached.

He suggested that until the danger from dust is reduced, shot firing should be either discontinued entirely, or that fire bosses withdraw the men from the mine at the time shots were fired.

Carpenter commented that in keeping with the recommendation by the coroner's jury, searches should be made of underground employees at stated intervals for matches, pipes, and tobacco, and added that it would be impossible to exercise too great care in such a matter. He, too, had his suspicions.

Carpenter also urged that the requirement for a ventilation plan be inserted into the Coal Mines Act. He also urged a clear definition of the terms "district" and "split," which had been the source of differences of opinion at the commission, but which would determine how many men could be put to work there.

The Hillcrest disaster also prompted some minor changes to mining regulations in Alberta, which were passed by an order in council. One new regulation prohibited any person other than an authorized individual from relighting a safety lamp underground. Another made provision for better regulation of the amount of coal dust in mines. The change with respect to safety lamps is curious, for it, at least superficially, seems irrelevant to the Hillcrest disaster. The Wolf lamps used at Hillcrest were self-lighting, locked safety lamps. No one needed a match to light them.

Perhaps more significantly, the province invested in mining education and increased training in mine rescue work. The province also began, largely as a result of the Hillcrest disaster, to organize classes in technical education at the mining camps. It should be noted that in addition to the 189 men killed at Hillcrest in 1914, nineteen other men were killed in underground mining accidents in Alberta during the same year.

On July 25th, 1914, a pit committee examined the mine and found that management had improved the ventilation with the new arrangement of four splits, but they also found a serious problem that raised questions about the competence of some fire bosses. They found gas in the no. 31 blind end on no. 1 south, and in what was known as Old 35. The discovery of gas was not unusual. What was of pressing concern, however, was that the gas had not been reported by any fire boss, despite daily examinations. The pit committee also found that dust had not yet been removed, nor had it been wetted down, despite a promise to do so. Thomas Thompson and Dan Howcroft of the pit committee concluded their report, "I might add we exonerate the Superintendent Mr. R.T. Stewart also the Fire Boss Hays as they are both new arrivals and have not in our opinion been in any way negligent."

The implications of the pit committee's report were so serious that they prompted a personal response from Stirling to the district mines inspector, James Richards. Stirling ordered Richards to give his immediate attention to having the gas cleared away, or to take steps to ensure that no men worked in the areas where gas had accumulated. Stirling pointed out as well that the fact that the men were new arrivals was no excuse for allowing gas to accumulate in the mine. The tone of Stirling's letter was unmistakably firm. The lack of proper inspection by fire boss Hays ultimately reflected on Stewart's management.

On February 23, 1915, less than a year after the disaster, the management of Hillcrest Collieries conducted a search for matches and smoking materials of all of the underground workmen in the mine. Such searches were recommended, although not mandated, by the Mines Act. No record was kept of whether such a search had ever been conducted at Hillcrest Collieries until this one. One man was found to have old cigar ends, but claimed that he only chewed them. Another man was found with a pipe full of smoking tobacco, but he fled the mine. A third was found with two matches in his possession, but claimed that he had left them in his clothes by mistake. District Inspector Shaw said he believed that the men found with tobacco did use it to chew, but suggested nonetheless that the men be prosecuted for their violation of the Act.

On March 6, 1916, Shaw inspected no. 2 north, but found the ventilation inadequate and men at work despite written and verbal instructions regarding the ventilation. By now, the provincial government had dropped all tolerance of mine safety violations in the wake of the disaster, and this was the final stimulus to action: he informed Stewart that he would be prosecuted for his failure to provide adequate ventilation in the mine.

While the commission of inquiry looked back at the cause of the disaster, others had become preoccupied with both the present and the future of the families left suddenly without breadwinners. In Edmonton on July 3, the second day of the inquiry, the Province of Alberta appointed a permanent commission called the Hillcrest Relief Commission, and granted it $20,000 for the relief of widows and orphans left dependent as a result of the Hillcrest disaster. All other moneys for relief would go into the same commission, including the $50,000 granted by the federal government. The district court judge for the Judicial District of MacLeod was appointed chairman, and the manager of the Union Bank, the treasurer. At the end of each year, a statement of the monies received and disbursed was to be published in the *Alberta Gazette*.

Edward Peel McNeill, an Irishman born in Dublin in 1863, assumed control of the relief commission, but as a judge at the District Court of MacLeod at Fort MacLeod, he also supervised the negotiations between the UMWA's District 18 and Hillcrest Collieries for the compensation that would be paid to the families of victims.

McNeill was a Freemason, as well as the 35th member of the Law Society of Alberta. Like Carpenter, he had earned his B.A. from the University of Toronto and his LL.B. at Osgoode Hall Law School. He had arrived in MacLeod in 1899 to practice law, which he did within a partnership firm at Fort MacLeod in the company of Frederick W.G. Haultain and Malcolm Mackenzie. In 1905, when Haultain departed, the partnership dissolved and McNeill practiced alone until 1907. He was named King's Counsel (KC) on March 12, 1913, and appointed to the District Court of MacLeod at Fort MacLeod one week later. He served on the bench there until 1921.

The last and by far the slowest of the money coming to families was from Workmen's Compensation. Under the Alberta Workmen's Compensation Act, which came into effect on January 1,

1909, employers were liable to compensate workers for injuries sustained on the job, and the dependents of workers who were killed at work. Claims had to be made within six months of the accident or the date of death, unless a reasonable cause could be given for the delay, and the employer was not liable if the injury or death was caused by personal negligence or willful misconduct.

If a worker was killed on the job, the employer paid to his dependents an amount based on the worker's previous earnings, or $1,000, whichever sum was larger, but not exceeding $1,800. For those only partially dependent on the worker, the amount changed. If there were no dependents, reasonable expenses of medical attention and burial were paid, up to $200.

All of these subtleties had to be negotiated between representatives of the union and the company, overseen by McNeill. They started the Hillcrest negotiations on December 11, 1914, and both sides signed the agreement on January 22, 1915. The indebtedness came to approximately $250,000, but would be paid by the company in installments. The company had already paid for all of the funeral expenses, so upon approval from McNeill, these costs would be subtracted from the amount already paid out. The company agreed to pay $3,000 every month to the office of the clerk of the court until the liability was fully paid, in roughly seven years.

The signed agreement provided payment of the full $1,800 to the collective dependents of each miner, not to each dependent individually. Each widow would receive $20 per month, and each child $5 per month, until the total money paid equaled $1,800. In the case of the thirty-two Austrians who were killed, no money would be paid to their dependents overseas until peace was declared between Austria and Great Britain. In the end, approximately ninety families were paid the full compensation. In 2012 dollars, the $1,800 amounted to approximately $36,594 per family, which, while better than nothing, was completely inadequate for women with no financial means to raise families.

Some women and children found support in their extended families. In many cases they were forced to share housing with relatives. Others applied to the general relief fund, and many moved out of the Crowsnest Pass, to where employment opportunities were better. Some were lucky enough to remarry. Employment opportunities for women were particularly scarce, however, with the result that many wound up with low-paying work cleaning homes or offices in Calgary or Lethbridge, sewing, cooking, or dressmaking. The widow of David Murray was left with twelve children to support, including her grandson, whose father was also killed in the explosion, and whose mother died in childbirth. After losing three sons and a husband at Hillcrest and another in the First World War, Elizabeth Murray was forced to move to Calgary with her ten children, where she worked for very little money as a cleaning woman in the afternoons and evenings. She never remarried. One is left in awe at the courage and strength of such a human being.

The disaster left the widow of the mine superintendent, James Sommerville Quigley, with five children and no husband to support her. Annie Barbara (née Maxwell) married James on December 24, 1895 in Westville, Nova Scotia. She was nineteen, he twenty-one. After the disaster, Barbara moved to Calgary, where she bought a rooming house to support herself and her children. She also worked as a seamstress to supplement her income. A staunch Presbyterian who attended church regularly, she never remarried, and passed away in Calgary in 1949.[1]

Mrs. Petrie, a widow, lost three sons in the disaster and a fourth to the First World War. She died in 1925 at the relatively young age of fifty-four. Later, on May 19, 1939, a fifth son would also die from injuries at Hillcrest when he was crushed by mine cars while performing his job as driver. A coroner's inquest failed to determine the circumstances of the accident.

George Hicken's widow Lily and their child were left with nowhere to go but to Lily's mother. Her sorrow at the death of her husband, her loneliness, and her financial difficulties became too much for her to bear. She died in 1923 at the age of thirty-two, as family lore has it, of a broken heart.[2]

The disaster also cut Mary Ann White deeply. She had arrived in Hillcrest in 1910 from Staffordshire, England with her daughters Elizabeth, Lily, and Mary and sons Alfred and Lambert. The other sons of the twice-widowed woman, Sam Charlton and Harry and William White, had emigrated first and found employment in the mines of the Pass. One daughter remained for a time in England, where she married Leonard Clarke; then they also emigrated. Lily married George Hicken. Elizabeth married the accountant at Hillcrest Collieries. In the disaster Mrs. White lost a son, Sam Charlton, and two sons-in-law, Leonard Clarke and George Hicken.

Many of the mining families were linked by marriage, perhaps because of the insularity of mining communities, or the bonds of common experience and values. Whatever the reason, the common responsibility of these men for each other's families comes into stark relief when one examines the aftermath of the Hillcrest disaster. Andrew Petrie, who had been working on a farm when his three brothers were killed in the disaster, and who was scheduled to work in the mine the next day, married Kate Anderson, the widow of another miner killed in the explosion. Another of Andrew's brothers, John, married Ellen Murray, of the same family that had lost a father and three sons in the explosion. Husband and wife had each lost three brothers. Like many others, John became ill from the conditions in the trenches of the First World War, and had to be hospitalized — but the hospital itself was mustard gassed by the Germans. It took John a year to die. In total, his widow Ellen lost a husband, her father, three brothers, and three brothers-in-law.

William Hutchison, Hillcrest's chief engineer, married the widow of pit boss Thomas Taylor, who died in the disaster. She brought four children to the marriage, whom Hutchison raised as his own.

These sorts of relationships were by no means unusual. The Ironmonger brothers, the Whites, George Hicken, and Sam Charleton were all related by birth or marriage.

When French immigrants François Labonne and his three brothers-in-law all died in the explosion, Labonne's widow Aglae was left alone to raise their two children, Martha and Frank. With no one to help her, like so many others widows of the Hillcrest disaster, she remarried. The man who assumed the responsibility for the family was another miner, Joe Toppan of Blairmore. In the fall of 1917 they left Hillcrest for good to take up farming at Three Hills. Another brother-in-law, August Geblin, married to one of Labonne's sisters, also worked in the mines, but lived to tell the story.

With the mine closed as a result of the disaster, and no certainty of its ever reopening, George Frolick needed a job.[3] He left Hillcrest for Calgary, but could not find work there, so he pushed on to Banff, where the CPR needed construction workers to help build a hotel. The man responsible for hiring, however, in a petty corruption scheme, demanded a $20 fee from anyone who wanted a job, and Frolick decided not to part with his hard-earned cash to get one. At this point, fate intervened when he received a letter from Brown, who asked him to return to work at Hillcrest. Frolick rounded up some friends and returned to the mine, where Brown thanked him for his work before and after the disaster, and asked him to become a driver boss. Though reluctant at first, Frolick finally agreed. He found the new job easier and got along well with the new superintendent, who had worked at the Michel mine until it closed due to high levels of methane. As a driver boss, Frolick took charge of the teams of men and horses that hauled the coal from the miners to the slants.

In May 1915, Frolick suffered an injury that would end his underground career: cars higher on the slant rolled backward and struck the cars he was standing between, crushing his leg. The Hillcrest physician, Dr. Rose, told Frolick that the limb ought to be amputated, but Frolick refused. Rose, who was a general practitioner, not a surgeon, reached for a medical textbook, studied it, and then, with the book open beside him, conducted surgery on Frolick's crushed leg to repair the damage as best he could. Using surgical nails, he fastened the top part of the leg to the pelvis. This procedure saved Frolick's leg, and with such success that in later years, other physicians would examine Frolick's leg and ask with admiration who had done the surgery. After Frolick's recovery, the management at Hillcrest gave him work in the lamp house, where he remained for many years. He and his family eventually returned to Ukraine, but returned to Canada, to Toronto, in 1939. Frolick passed away in 1982.

John Brown's days as general manager of Hillcrest Collieries came to an end discreetly. Perhaps it was his musings in the night about what he might have done differently to prevent the disaster, or else a polite suggestion by the board of directors that it might be a good idea to move on, that led him to make the decision about his future at Hillcrest. In any case, as *The District Ledger* reported on March 6, less than a year after the last of the bodies had been buried, and while Brown was away, Frank Drinnan arrived at Hillcrest to take up the duties of general manager. Brown eased Drinnan through the transition, and then departed for his new home in the United States sometime in April. A crowd of friends saw his family off at the station. Now, no one would be reminded about the disaster when they spoke to the general manager, nor wonder if he was, in some indirect, indefinable way, responsible. The cobwebs had been swept clean. One can only wonder: if the board had hired the more educated Drinnan in the first place, would the disaster have been prevented?

The First World War brought a new set of problems to the Pass, and a particular problem for the UMWA and mine management. On June 15, 1915, the British and Italian miners at Hillcrest refused to work with the German and Austrian employees after the Hillcrest manager had refused to fire them. By this time, about seventy-five Ukrainians, citizens of the Austro-Hungarian Empire, worked at Hillcrest.

Many of the men from the Pass had volunteered for the Canadian Expeditionary Force. Casualty lists of the dead, and missing and wounded soldiers from the Pass, along with reports of Allied victories and losses, appeared regularly in the local newspapers, and sentiments against Germans and Austrians ran high.

The thrust for the stand taken by the British and Italians may have come in part from Bellevue, where the company laid off a large number of men in mid-May, because, they said, they wanted to divide up work, which was scarce. Those laid off, however, were all Austrians and Germans, and their "temporary suspension" had extended to mid-July. *The Ledger* commented sarcastically that "possibly the Austrians and Germans at Bellevue are going to 'change about' every year or every decade with those now employed."

According to the British miners at Hillcrest, threats had been uttered by Germans or Austrians, and some of the Brits felt that their own safety would be threatened if they were to continue working with alien enemies.

The decision not to work had been made at a morning meeting of employees at the pit head. A so-called citizens' committee had then been formed and began wire negotiations with officials in Ottawa to have the Austrians and Germans interned, as had been done at Fernie. The Hillcrest miners, in their walkout, had mimicked the behaviour of miners at Coal Creek in B.C., who had done the same thing, walking off the job despite warnings from UMWA officials. The *District Ledger* was quick to point out, in a June 19 article about the internments at Fernie, that the actions at Hillcrest were initiated by individuals, and not the UMWA. Almost a year to the day after Italians and Brits had stood side by side with German and Austrian miners in the rescue work, sentiment aroused by the war had led to discord within the union ranks.

About 150 Germans and Austrians worked at Hillcrest, many of them married with children. Clearly, those with dependents could not be interned. Even federal officials could not bring themselves to intern family men. The walkout held up nearly the mine's entire operation, although management had nothing to do with the dispute. Eventually the British miners were satisfied that the other miners' comments had been inaccurately reported, but the foreign workers were suspended anyway, until the government decided what to do with them.

On the evening of the 17th, the Hillcrest miners agreed to resume work on the 19th, the one-year anniversary of the disaster. The tension, though, had not evaporated. The miners did not object to working with the few Austrians employed in haulage, as well as others needed to complete particular jobs, but no foreign coal diggers were allowed. The miners agreed that their places would not be filled until the whole question was settled. The new Hillcrest manager promised that if charges could be proven to be in the slightest way dangerous, he would summarily dismiss the guilty individual. This seemed to calm the protestors' suspicions. Drinnan asked George Frolick to round up all of the Ukrainians and to go to work immediately. Frolick and the others understood what was at stake, and worked as they had never worked before, each doing the work of two or even three men without complaint, while the British looked on in amazement. Because the strike was illegal, no one could, in good conscience, call the Ukrainians scabs. After watching them work for a time, the strikers became bored and returned to work.

Meanwhile, the German and Austrian men imprisoned in the skating rink at Fernie, who were allowed to keep a flute, an accordion, and a melodeon, played "It's a Long Way to Tipperary" over and over again as friends brought them cigarettes, day dissolved into night, and days into weeks.[4] Their denials of disloyalty to their adopted country had been ignored, and only through music could they discharge their frustration, or express their loyalty to

Canada. No bedding was provided. The men slept on the concrete floor and were not provided with a means to bathe.

Frolick, although an immigrant from the Austro-Hungarian Empire, was not imprisoned, as he was a Canadian citizen. He was obligated, however, to carry a special identity card for the duration of the war and expected to produce it whenever he was stopped by police.

Perhaps it was the spirits of the miners who had achieved brotherhood in death who reminded those alive of the right path. On the 19th, the anniversary of the disaster, about a thousand people gathered at the Hillcrest cemetery to remember the victims. The Oddfellows gathered in their hall in Bellevue and 120 paraded to Hillcrest, led by the Coleman city band. Later, a light supper was served at the union hall, followed by a memorial service at 7:30 PM, with addresses delivered by Reverend Brother J.F. Hunter of Blairmore, Reverend T. Hart of Hillcrest, and Calgary MLA R.E. Campbell. Hart, who no doubt thought of the ill feelings between miners as a result of the war when he spoke, addressed his audience on the subject of "Disaster and Brotherhood."

At the internment camp near Fernie, Lieutenant Colonel Joseph McKay had taken over the prisoners from the provincial police, and began to sift through the cases of the roughly 330 men interned to determine, case by case, if any should be released. The first to be examined was John Dendis of Coleman, who was released on July 10 after it was found that Dendis had a wife and three small children dependent upon him for support. Another twenty-four men were released on July 16. By the 19th, McKay had released 157 labourers and lumbermen who willingly took the mandated oath, and signed a pledge to report when required to do so by the police. All were young, many under the age of twenty-one; one was only seventeen.

As well, McKay had discovered mistakes made by the arresting officers. Among those arrested were an American, eight Bohemians, and an Italian, all of whom were released, presumably with apologies. One interned Austrian who could have been released said he loved Canada, but could not find a job, and did not want to leave the detention camp. The authorities saw fit to cooperate, and the man remained in detention, where he was provided with meals and a roof over his head.

In the end, only two of the twenty locals in District 18 agitated for internment: Fernie and Hillcrest. The miners at Coleman met to consider the question, but decided to take no action. For the union, it was an embarrassment that they addressed in a strongly worded front-page editorial in the *District Ledger* on July 10, 1915. The newspaper, which identified itself on its front page as the official organ of the UMWA's District 18, titled the piece, "The Situation in District 18, As Viewed by District Officers." It was a pragmatic call, however, not for the principle of justice, but for union solidarity. The editorial called upon union members to recognize the truly international makeup of its membership, and reminded them that they had no qualms accepting men of different nationalities as brothers, neighbors, friends, and working partners. The writers suggested that as union men they had joined hands for mutual protection and effective collective bargaining, and they could not now reject their fellow workers. Almost as an afterthought, they commented that the men were not lawbreakers. Many of the internees, the writers pointed out, had now, as a result of the miner-initiated internment, angrily rejected the union. This was at the heart of why the union objected to the internments. The writers further pointed out that the British had not benefited from the internments, as the vast majority of the replacement workers were of two other foreign-speaking nationalities. The article concluded, not with an appeal to recognize the injustice of the internments, but to reject them in the interest of union solidarity.

When war was declared in August 1914, William Hutchison still worked as assistant general manager and chief engineer at Hill-crest. He was almost forty years old, but wanted badly to serve. He tried signing up with the Princess Patricia's Canadian Light Infantry in Calgary, but found that he had applied too late, so he signed up instead with the militia unit in Pincher Creek. When he didn't get the call from the militia into active service, he took it upon himself to begin recruiting a new battalion that would consist of men from Hillcrest, Frank, Blairmore, and Coleman. It would become the 192nd. Hutchison was put in charge of prelim-inary training, and set up his headquarters at the old Sanitorium Hotel at Frank. From there they were moved to Sarcee at Calgary. Hutchison and another man were then sent to Rockcliff Camp in Ottawa to take a machine gun course. That task completed, they returned to their own battalion at Sarcee, where Hutchison passed the examinations for lieutenant and captain's ranks. They were shipped to England for more training, but Hutchison became fed up with delays when he was told that he was too old, and so man-aged to get himself transferred to the Canadian Forestry Corps (CFC). Finally, he was ordered to a location on the Arras-Bapaume Front in France, where he was given responsibility for the fell-ing and sawing up of trees that had been strafed. Eventually he was given responsibility for the construction and management of sawmills that provided lumber for railway ties, trenches, bar-racks, hospitals, road surfaces, and crates for ammunition. The CFC produced seventy percent of the Allied timber used during the war, and Hutchison made an invaluable contribution to the war effort.

David Hutchison remained as surveyor for the mine through the war and beyond, until it closed in 1939. He was rejected for service in the First World War because of poor vision in one eye. As mine surveyor, he mapped all of the workings and drafted blue-prints in the mine office. In 1918 he married a teacher, Isabella MacLeod, who had moved west from Cape Breton. They raised two children. David stayed on with the mine into 1940 to update the mine blueprints, as required by law, then worked surveying new air fields for the Commonwealth Air Training Plan during the Second World War. He died in Edmonton in 1970, at the age of eighty-nine.[5]

Albert May, who had been the outside engineer at the time of the Hillcrest disaster, and who had missed serious injury or death by no more than ten minutes during his inspection of the no. 1 fan, joined the 192nd battalion in January 1916, and saw action in Passchendaele, the Third Battle of Ypres, in 1917. In the rain, mud, and poor sanitation of the trenches he contracted rheumatoid fever, which damaged his heart. This led to his hospitalization in France, England, Quebec, and finally at the Frank Sanatorium. Although he never fully recovered from the effects of the war, he married in 1921 and continued to live in Blairmore, an active member of the Masonic Lodge until he passed away in 1945.

George Wilde, who escaped from the mine after the disaster, then assisted in rescue efforts, signed up to join the Canadian Army in Pincher Creek in February 1915. He was shipped overseas, where he fought in the trenches and was killed in battle. He left behind a widow.

Joe Fumigalli, who helped Jack Eddy transport the bodies from the wash house to the town, bought the draying business from Eddy later that year, including a livery barn with horses, a three-seat Democrat buggy, sleighs, and wagons.[6] An energetic worker, he snagged the contract from the CPR to haul freight, express, and mail from Hillcrest station to Hillcrest town, and hauled wood and coal from the mine to homes in the town. Fumigalli bought one of the first cars in the Pass, a Model T Ford, then used it to operate a taxi service through the Pass. In 1917 he built a garage on Main Street in Hillcrest, and obtained the Oldsmobile dealership. He sold his first car to Bill Hutchison.

Upon his return from the war, William Hutchison resumed his duties at Hillcrest Collieries, and married Elizabeth Taylor, the widow of pit boss Thomas Taylor. Hutchison saw the mine through expansion and development during the turbulent 1920s.

During the postwar years, through the 1920s and into the 1930s, internecine feuding consumed organized labour in general, and the UMWA in the Crowsnest Pass in particular. As a result of the struggles among the various factions of union radicalism, passions became inflamed, allegiances by miners altered, labour unions appeared, evolved, and faded away, and strikes came and went. Through it all, the lives of the people of the Crowsnest Pass went on through bad times and good. The old died, women gave birth to babies, and miners dug coal.

The widows of the disaster and their children struggled on with whatever assistance was available. The statement of the Hillcrest Relief Commission for the year 1918 reveals that the balance in the fund totalled $33,206. The expenditures for the year totalled less than $5,000. Receipts totaled $870. The fund manager had left the only liquid assets of the fund, $7,896, in an account at the Union Bank of Canada in Bellevue. He had placed the rest in Victory Bonds and Province of Alberta Savings Certificates.

By 1919, labour in Canada had begun an internal struggle to determine its direction, with a pronounced difference of opinion between Eastern and Western Canada, between international craft unions and radical industrial unions. The outcome of the ensuing conventions was the official formation of the radical anti-Capitalism One Big Union (OBU) in Calgary in March, 1919. District 18 founder Frank Sherman's son Willy, described by the RNWMP as a communist, was one of the militants within the OBU at Fernie in 1919 and 1920. The District 18 miners voted to join the OBU and withdraw from the UMWA in part because of the UMWA's loyalty to the war effort, which they opposed. The first OBU convention called for a general strike to back the demand for a six-hour workday and a five-day work week. By mid-May the OBU was running District 18, and was using UMWA money to finance its operations; however, the use of UMWA money only fuelled the UMWA's resistance, which gained the support of the coal operators as a less radical alternative to the OBU. The operators refused to bargain with the OBU, the UMWA gained pay increases for miners, and District 18 and the UMWA resumed their relationship August 1, 1920.

The power struggle between the UMWA and OBU spilled over, however, into the victims of the Hillcrest disaster. In January 1920, UMWA officials published a circular that reported the results of an investigation by a chartered accountant into the accounts of District 18 while it was under the control of the OBU. In particular, the report dealt with transactions relative to the Hillcrest Widows' and Orphans' Fund. The chartered accountants found that $548.30 was missing from the fund, which had been controlled by former District 18 president Phillip M. Christophers and secretary-treasurer Ed Browne, both of the OBU. Christophers would later reject association with the OBU, sit in the Alberta legislature as the member for the constituency of Rocky Mountain (which included the Crowsnest Pass), and rejoin the UMWA. The auditors demanded repayment, and received a cheque from Browne. But the auditors also found that $1,870 had been transferred from the Relief Fund to the General Revenue Fund and was never paid back.[7] To make matters worse, when the UMWA resumed control, they found the general fund depleted, with many outstanding debts against the district. The UMWA termed the actions of the OBU a breach of trust. Once again, victims of the Hillcrest disaster had been all but forgotten in the struggle for power and influence.

By 1920 Hillcrest Collieries mined an average of 1,300 tons of bituminous steam coal daily, and employed an average of 425 miners. The population of the town stood at 1,200, and boasted a Methodist church, a public school, local and long-distance telephone service, a movie theatre, and two meat markets. The Cruickshank-Burnett store, in addition to its two owners, now employed four staff, and had competition from a dry goods, boots, and shoe store. James Gorton became a vendor of soft drinks and cigars at the Great War Veterans Association, which also operated a pool hall.

The First World War had begun to change how people related to each other in slow but noticeable ways. Racial and ethnic tolerance grew, and certain ethnic groups could cross the cultural and career barriers that had restricted pre-war immigrants. It became incredibly difficult for one man to ostracize another based on his ethnicity when they have fought together, bled together, and watched their friends die together. Tony Casagrande became the Hillcrest undertaker, as well as the yard foreman. Fred Hrihoriw now owned the Union Hotel. Jim Chan ran one restaurant, Jim Quong another restaurant and confectionary. Sing Kong ran one laundry and Sing Sam another. Tony Bovio became a pit boss, as did John Ironmonger, the survivor of the 1914 disaster. Rinaldo Fumagally ran the livery. Percy Salt ran a market garden.

On January 6, 1923, Trenholme Dickson, a solicitor in the attorney general's department, wrote a letter to the acting clerk of the court at MacLeod, Lillie Thomas, to ask for information about a widow who had died recently, leaving behind three underage children. The Hillcrest disaster had widowed Mary Kane. On January 8, the Alberta government's superintendent of dependent and delinquent children sent another letter, this one to the office of the official guardian, asking if the family would be entitled to Hillcrest relief, and stated that the older sister of the children was very anxious that something be done for them. On January 12, Thomas replied to Dickson. She said that she had noticed Kane's obituary in the *Lethbridge Herald,* and she knew that there were children left. Thomas wrote to Father McCormick, the man who buried Kane, to find out what position the children were in and whether they intended to carry on the home. The only relief extended to Kane by the Hillcrest Relief Commission was a supply of coal and light.

Father McCormick informed Thomas that five children remained. The children wished to keep the home going and to stay together. There appeared to be some debts against the family for articles of furniture, and also accounts with the grocer and butcher. The family's income was generated solely by the two eldest children, who had jobs.

Thomas wrote that she had submitted the information to Judge MacDonald, the chairman of the Hillcrest Relief Commission, who suggested that an allowance of five dollars per month be made to each of the three younger children and that the amount be forwarded to Kane. This was in addition to the $1,800.00, less $57.80 for funeral expenses, she had already received as compensation for the death of her husband under the Workman's Compensation Act. The money was paid in installments, ending in June 1918. After the compensation was exhausted, she was placed on the relief list and received $1,098.30 from the Hillcrest Relief Commission, but this assistance was discontinued in April 1920. From then on she received free coal and light.

Thomas reminded Dickson that "there is no amount to which any party is entitled from the Relief Fund, but allowances are made to various parties from time to time as the Commission considers necessary."

While the families of the Hillcrest victims struggled to survive, the District 18 infighting continued. Reality began to conflict with what communist-inspired union leaders believed miners were entitled to for their labour. As petroleum became more popular, the use of coal declined, but there were broader issues at play. After the war, the demand for copper plummeted, and copper production consumed coke. Union leaders forgot, or never knew, or refused to accept the fact that everything changes. The huge industry that involved horses as the mode of transport, for instance, altered and shrank, and the men whose lives had centred upon horses moved on. If there were calls to fight the drop in wages in the horse industry, they were dismissed as outright nonsense. Wages rose for mechanics and those in auto assembly plants just as they dropped for blacksmiths, harness makers, and saddlers. The fact was that too many miners chased too few jobs, a result of a lowered demand for coal, a substance as common as dirt. The facts were simply an invitation for miners to move on.

In 1924, John Blue summarized the situation in the Alberta coal fields. "At present the mines are operated little more than half the time, resulting in dissatisfaction among the workers and increasing the cost of production. Shortage of railway cars, and misrepresentation of the size and quality of coal have led to a restriction of the market in places where Alberta coal comes into competition with American coal." He added that 600 mines had been opened since 1905, but only 276 were still in operation. Put another way, sixty-four percent had been abandoned. In the last 15 years, Blue wrote, every 100,000 tons of coal produced had cost a human life.[8]

While the influence of the UMWA waned, the Western Coal Operators Association collapsed entirely on November 28, 1925. Smaller company associations replaced the big unions, and the Hillcrest miners turned their support to the British Columbia Miners Association along with Fernie and Michel. When the Communist Party of Canada attempted to organize all of the associations on June 1, 1925 at Blairmore, miners from Hillcrest and Michel attended, but declined to involve themselves with the new Mine Workers Union of Canada, which included associations from Blairmore, Coleman, Bellevue, and Corbin. A large segment of miners did not favour the Mine Workers Union of Canada, the UMWA, or unions in general. The Hillcrest miners stood among their ranks. When the Great Depression descended like a dark curtain in the fall of 1929, however, demand for coal began to drop off even more precipitously, and with the worsening economy of the coal fields, radicalism among miners increased. The crash slashed rail traffic in half, the production of steam coal plummeted, and massive unemployment followed. In 1930, the Great Northern Railroad cut back on orders for steam coal because of its switch to fuel oil, a decision that dealt a huge blow to the mines at Coal Creek and Michel. The Mine Workers Union of Canada was no longer viewed as radical enough for miners, who shifted their support to the Workers Unity League of Canada, a communist labour organization, and demanded the radicalization of the Mine Workers Union. By August of 1930, communists controlled it.[9]

In early 1932, the union drew up a list of demands that included a ten percent wage increase, a six-hour workday, no night shifts, and time and a half for overtime. The mine owners, as one might expect, rejected the demands. The strike that ensued involved every miner in the Pass except those who worked at Hillcrest Collieries. At Bellevue, West Canadian Collieries announced that it would reopen the mine on May 4. A mob assembled at the mine, determined to prevent its opening, but was met with seventy-five heavily armed members of the RCMP (formerly the RNWMP).[10] While the mob tossed pepper and stones, the policemen fought back with batons. During the hand-to-hand combat, which ensued for two days, the police arrested more than a dozen strikers, but the Bellevue mine closed indefinitely in the face of the violence.

In mid-August, Alberta Premier J.E. Brownlee travelled to the Pass and urged both sides to talk. The result was an agreement signed on September 6, 1932, which the union claimed as a victory even though they gained neither their wage increase nor the workday they wanted. The mines reopened to an uneasy peace.

On the B.C. end of the Pass, the mines began to die. Coal Creek closed, and at Fernie, management shut down the coke ovens in 1933. At Corbin, during a bitter and violent strike in 1935 that involved a battle with police, mass demonstrations, and a rejection of a conciliation board by the miners, the owners closed the mine permanently on May 7. The communist-run union called it a bluff. Only when the rails of the Eastern B.C. Railway, Corbin's link to the outside world, were torn up and hauled away did the stunned miners and their families get the message. Subsequently, eleven men involved in the riot, which resulted in injuries to policemen and miners alike, were fined and sentenced to six months in jail. The closure of the mine at the hands of the militant Mine Workers Union of Canada, and the stiff jail terms handed to the rioters, contributed to the radical union's collapse. Even the Communist Party of Canada that controlled it had to face a loss of credibility. A spokesman urged miners to rejoin the UMWA, which they did.

The miners and their families of the Crowsnest Pass, like most Canadians, faced enormous financial pressures during the Great Depression. Miners shared what work there was in the mines, and workers at Hillcrest would wait anxiously for the mine whistle to blow at 8 PM, which meant they were needed to work the next day. A man who got work at the Hillcrest tipple in the 1930s would earn $4.45 per day. In a good year through the 1930s, a miner could earn up to $1,500, but only three out of ten miners worked 200–300 shifts per year. Even by 1938, coal production in the Crowsnest Pass had only returned to sixty-five percent of its former levels. The vegetable gardens and livestock kept by the families of miners often made the difference between hunger and a full belly. But the families persevered, and even found happiness in the towns of the Pass, which had known so much sorrow.

On September 19, 1926, another massive explosion rocked Hillcrest Collieries, but killed only two men: fifty-five-year-old fire boss Frank Lote and twenty-five-year-old pump mechanic Fred Jones, the only men in the mine at the time. Fire boss Dan Kyle, who had escaped death on the *Titanic* and in the 1914 disaster, had been scheduled to work but traded shifts with Lote. Kyle's extraordinary luck saved his life a third time. This explosion was worse in terms of destructive force and size than the 1914 blast.

The explosion of 1926 appears to have been the real thing: one of the rare circumstances in which sparks created during a fall of rock ignited residual methane. For once the source of ignition was clear. As a result of this blast, the provincial government imposed new regulations mandating that all mines in the Crowsnest Pass had to be dusted with rock dust to help prevent coal dust explosions.

By 1935, Hillcrest Collieries habitually lost money, and in the winter of that year the directors met in an emergency session in Montreal to discuss what to do about it. Among those in attendance were Sir Charles Gordon of Dominion Textiles, and William Hutchison, whom the directors closely questioned and asked for suggestions. He recommended that coal only be mined in districts of the mine where the cost of extraction was lowest, that sales of commercial coal be cut back, especially on sizes that were expensive to produce, and that an effort should be made to get a bigger share of the supply of CPR coal. At the conclusion of the meeting, they appointed Hutchison general manager of the company.

Under Hutchison's management, the Hillcrest mine merged with the Mohawk mine to form a new company known as Hillcrest-Mohawk Collieries Ltd. He remained for two and a half more years as general manager while the new company began operations. The company made money during this period, but Hutchison left for Vancouver in 1939 to be with his ill wife. During the Second World War, the company hired him back as a consultant, to help them find lower ash coal needed to meet increased wartime demands. Hutchison had explored the area thoroughly in the early days, so he travelled to Hillcrest and showed them the location at Byron Creek. A short while later they called him out again to design buildings for the new Byron Creek mine. He stayed to complete the design work, and left when the construction began. For a time he lived at West Bay in West Vancouver, until 1950, when he moved to Penticton.

In the last line of his memoirs, Hutchison wrote, "It is now 1959. I am in my eighty fifth year and still active." Hutchison passed away in 1965 at the age of ninety-one, and was buried in Vancouver. His legacy was that no man contributed as much to the plans, development, management, and success of Hillcrest Collieries from 1910, when Charles Plummer Hill vacated management, to its final closure in 1939.

CHAPTER 17
THE CLOSURE

On December 4, 1939, the Hillcrest Collieries liquidator informed Alberta's chief inspector of mines, A.A. Millar, that "[i]n conformity with the provisions of The Mine Act, we hereby advise you of our intention to abandon the Hillcrest Mine within the next ninety days." The next job for the liquidators was to salvage the mine equipment, which included the steel rails, the mine cars, the hoists and boilers, generators, tools, and anything else salvageable for use or sale elsewhere. Much of it would go to the Mohawk mine. By late December, the company employed thirty-one men to remove material, with the expectation that everything would be out of the mine by the end of January. As the district inspector of mines commented, "Hillcrest will no doubt become a ghost town in a very short time." He was wrong.

There is a certain irony to the identity of the victim of the last-reported accident at the Hillcrest mine. At 8 PM on November 7, 1939, a driver was hooking a tail chain to a car outside the no. 1 mine when his feet became tangled in the lines and the horse bolted, dragging him along the track. The mishap injured the man's left groin and bruised his right ankle. He was able to continue work, but finally had to undergo surgery for the hernia caused by the accident. The victim was twenty-five-year-old Sam Ironmonger. His injury could hardly be described as grave, but was certainly noteworthy in that he was a member of the same Ironmonger family that had suffered so grievously in the 1914 disaster, with the loss of brothers Charles and Sam.

If this second Sam Ironmonger had looked at the ground, close to the wall of the no.1 hoist house, he would have seen, partially buried after a quarter century, a large piece of angular, broken concrete. Likely he would not have known what it was or where it came from. Only then, with the buildings in ruin and the parts of the hoist house wall that were not made of concrete removed, could an observer see where the forgotten piece of concrete had come from. The piece on the ground had been blasted from the wall of the hoist house, along with the rest of the wall and the entire roof, by the force of the 1914 explosion. Like an incomplete jigsaw puzzle, the only place that the half-buried triangular piece will fit has been revealed, and now the broken wall and its broken piece sit, forever separated, a reminder of the lives broken by the Hillcrest disaster of June 19, 1914.

The union had done its best to prevent the closure, to no effect. The Hillcrest local held a special meeting on December 2, which was also attended by the Mohawk mine pit committee. Sadness filled the room as each man contemplated his future, its uncertainties, and the inevitable fractures in longtime friendships that would come with displacement. They voted first to open the meeting to anyone who wished to attend, including reporters. The president, E. Rees, outlined the events that had led to the closure of the mine, and the steps that had been taken to try to prevent it. They had made a good case, but legal and other obstacles had gotten in the way. The coal was owned by Hillcrest Collieries and the government felt it had no power to enforce operation.

The union had tried to get permission for the workers to mine coal for themselves, but that also proved practically impossible. Only a few details remained. The local wanted to ensure that the Hillcrest miners had jobs, but could not offer them employment at Mohawk until all of the laid-off Mohawk miners were reinstated first.

The men walked into the mine for the last time on December 4. They had worked their last shift on the 1st, the day before the meeting. This time they went inside not to pick up tools to begin the day's work, not to pick and shovel coal, set props, and lay steel, but instead to remove tools, and to bring out the mine car checks. It was the last part they would play in the history of the mine, and was heavy with symbolism for miners, who could not work without tools, and whose work was measured and pay calculated by the checks.

On February 13, 1940, the district inspector informed Chief Inspector Millar that "[a]ll material is now out of the mine, and they are now closing up the mine entrances, there are six in all, namely; #1, #2, and #3 slopes, also the entrances at #45 fan, Falls Creek fan and #1 fan." Each entry to the mine was sealed up with a cog made of sound timbers that were 12 feet long, and laid skin to skin for the width of the entry, and under the lip of the hill.

On April 10, liquidators informed Millar "that the Hillcrest Mine, Provincial No. 0040, has been abandoned, effective as at 31st. March 1940. All entrances to the underground workings have been properly sealed in accordance with the requirements of Section 100 of the Mines Act." The Hillcrest mine had closed.

The legacy of union men mistrusting military authorities and police, engendered by the radical unionism of Frank Sherman and others, found a place among the next generation of miners, but not in their hearts. As news of the impending war in Europe filtered into the Pass over the radio waves from Calgary, the teenage boys who gathered and hung out together swore not to volunteer or be conscripted. In their naiveté they imagined themselves hiding out for the duration of the war deep in the mountains. Frank Hosek, son of a Bellevue miner, grew up in the Pass. He says in his memoirs, "It may have been due to the life endured in a coal mining community where work was not always available or because of the war in Spain, but the feeling was, we did not owe the country our lives." When the war actually began, some few who sought adventure volunteered first. But soon, said Hosek, that all changed:

> "This was Bellevue, a pin prick of a place tucked away in a remote corner in the Province of Alberta. Here, where we lived, played, and went to school, we took only a scant interest in world affairs and yet news of the war and of the need for men to fight it filtered through our isolation and nibbled away bit by bit at our determination to stay aloof from all that was happening in the outer world. It soon became apparent that more and more of the young from this small corner of the world called the Crowsnest Pass and of which Bellevue was but one small town were answering to the call. The lure of adventure, perhaps even patriotism may have bid some to go, but unemployment and the dwindling numbers of those who stayed behind were the deciding factors for us all." [1]

And, as among the ranks of men from beyond the Pass, who stepped from the farms, forests, towns, and cities of the west (where the poverty and the dole had also broken men's spirits during the '30s) and into the recruiting offices, many would never return to their homes, their lives forfeited, as others had been in the first great conflict.

In the years that followed its closure, some residents of the town forgot where the mine had been. A few people wandered up the approach road to the mine — hunters and children from the town — but over the years the forest grew up again and reclaimed the land. Gradually, nature began to erase the traces of what had been there, and many of the men who had worked at Hillcrest Collieries and their families moved away, and grew old, and passed on.

The mine imparted a richness to the town of Hillcrest, not just in the money that passed through the calloused hands of the miners, but also in the lives it gave to its residents who were born, thrived or endured, and died within the circle of mountains enclosing the town's homes and streets, in the legacy of grit and determination left by the miners, and even by their all-too-human frailties. The mine lured immigrants from the varied palette of the European landscape and forced them to work together. Their descendants now stride through all walks of life, with the ethnic barriers dissolved into an amicable, agreeable homogeneity. They are Canadians now, and have no need for Ukrainian Halls or Italian Clubs. Jim Hutchison, the son of Scottish mine surveyor David Hutchison, chummed as a boy with Wing Chan, the son of a Chinese immigrant who ran the restaurant in Hillcrest, Lenin Gryschuk (whose brother was named Marx), the son of a Russian immigrant miner, and Harold Stefano, the son of an Italian miner. And so it was, and is, among all of the descendants of those brought to the mines of the Crowsnest Pass.

Of the men that the disaster brought briefly to Hillcrest in June and July of 1914, Judge Arthur Carpenter continued to work as a dedicated Alberta public servant. A year and a half after his work at the head of the inquiry, Carpenter resigned his position as district court judge in Calgary to sit on the board of the Public Utilities Commissioners in Edmonton. He accepted the chairmanship of the board, and remained there until retiring in 1939. He had previously been named King's Counsel (KC) on February 4, 1919. Members of the Calgary Bar Association held a banquet in his honour upon his retirement and presented him with an engraved silver plate. It remains in the possession of his descendants. After his retirement, Carpenter opened his own law office in Edmonton, but before long the war effort called for his skill and knowledge, and in January 1942 he was appointed regional enforcement counsel to work in connection with the Alberta office of the Wartime Prices and Trade Board. He later worked with the Alberta Mobilization Board. In 1943 he was appointed president of a new crown company, Wartime Oils, Ltd., organized to explore means of increasing oil production in the Turner Valley. Carpenter eventually married, and he and his wife had raised two daughters together. Carpenter died on February 17, 1949 at the age of seventy-six, his entire life dedicated to public service in the province he came to love dearly, and defended fiercely throughout his life.[2]

The career of W.F.W. Hancock, one of the RNWMP officers who had been rewarded by the prime minister for his work during the Hillcrest disaster, was also one of distinguished public service. He joined the British Army when the First World War began, and after service with the rifle brigade in France he joined the Alberta Provincial Police as an inspector at Peace River. Second-in-command of the Alberta force when it disbanded and amalgamated into the RCMP in 1932, Hancock rejoined the Mounties with the rank of inspector. In 1936 he became superintendent of "K" Division headquarters in Edmonton, and in 1937 became assistant commissioner, responsible for all of the force's police work in the province until his retirement in 1946. During the Second World War Hancock worked with American forces building the Alaska Highway. For that work he was awarded the Order of the British Empire. After the war, Hancock took on a public relations post with Northwest Airlines. Later, appointed an Alberta provincial judge, he sat on the bench in Stony Plain and Edmonton. In 1960, Magistrate Hancock was elected chairman of the board of the waterfowl habitat conservation organization, Ducks Unlimited, after serving as president. He retired to the West Coast in 1965, and passed away there in 1975 at the age of eighty-seven.

RNWMP Inspector Cortlandt Starnes and Corporal Frederick Mead continued their policing duties long after their involvement in the Hillcrest disaster. Starnes commanded the RNWMP in Winnipeg in 1919, at the time of the Winnipeg general strike and riots. On May 15, "fifty-four mounted men under Inspectors Proby and Mead, followed by thirty-six men in Motor Trucks" rode in to restore order. "Mead was the man whom labour officials had gratefully praised for washing and laying out the mangled bodies of the union men killed at the Hillcrest Colliery disaster, but all that was forgotten now." Inspector Proby was struck from behind and owed his life to another policeman who felled a rioter in the act of aiming a gun at the disabled officer. A constable was knocked down and beaten, then dragged by rescuers to safety into an undertaker's parlour, where the body of Major-General Sam Steele[3] awaited burial.[4]

During the First World War, Mead served overseas with the RNWMP Cavalry Draft. In the mid-1920s, he served at Lethbridge, was promoted to superintendent in 1933, and assumed command of "C" Division at Montreal. In 1936 he became assistant commissioner, in 1938 took command of "H" Division at Halifax, and in 1941 moved to RCMP headquarters in Ottawa. Mead was a member of the B.C. Security Commission, which arranged for the removal of Japanese from the coastal areas of B.C. during the Second World War.[5] In 1944 he was promoted to deputy commissioner, and retired in 1947. Mead was made a Commander of the British Empire for his service with the RCMP, and died in Vancouver at the age of seventy-two.

Starnes was appointed the 7th commissioner of the RCMP in 1923, a position he held until 1931. He died in 1934.

Christen Junget served nearly thirty-seven years with the NWMP, the RNWMP, and the RCMP. He was transferred to Blairmore in 1916, and to Fernie three years later. In 1922, he took command of "K" Division at Lethbridge, and that same year was promoted to superintendent. In 1935 he retired to Oak Bay, east of Victoria, where he lived in a house with a large living room window overlooking the Juan de Fuca Strait, with San Juan Island and Mount Baker visible in the distance. He retained only a trace of his Danish accent.[6] Junget died in Victoria in 1969.

George Cruickshank, whose Hillcrest store had supplied the coarse cloth in which the bodies of the dead miners were wrapped, and for which he was never paid, bought out his partner's share of the store in the late 1930s. During the many mine strikes over the years, Cruickshank continued to extend credit to hard-pressed miners, even into the Depression. Shortly before the 1930 provincial election, Cruickshank decided that he wanted to run for office. Voters in the Pass elected him to the legislature in 1930 for the Rocky Mountain constituency, and he sat as an independent during the tenure of the Brownlee UFA government until 1935. Cruickshank continued to operate his store in Hillcrest until his wife's death in 1948. From then on he lived with his daughter in Blairmore, and spent his winters in Arizona or Victoria. Cruickshank died of a heart attack in 1962, and was remembered above all as an honest man.

Charles Plummer Hill and his wife Enid never returned to Hillcrest. They moved to Montreal, where they lived for 11 years, and then, drawn again by the lure of the west, moved finally to Victoria in 1921. They lived out their last days in a grand house that they dubbed "Hillhaven." It looked out over a small cove at West Bay. Hill had done well financially when he sold the Hillcrest mine, and continued to do well with investments and the steady income earned by his stock ownership in Hillcrest Collieries. His life, in contrast with his earlier years in the west before he founded Hillcrest, when he often lived in tents or under the stars and travelled by horseback, had become one of leisure and relative ease. He retained an interest in mines, and became shareholder and director of several mine companies, and even held the position of president with MacLean Underfeed Stoker. He moved in the upper level of Canadian business circles as a lifetime member of the American Institute of Mining and Metallurgical Engineers, and as a member of the Mining Association of British Columbia and the Canadian Institute of Mining and Metallurgy. Socially, he held memberships in the Constitutional Club of London, England, the Vancouver Club, the Union of British Columbia Club, Victoria's Colwood Golf Club; he also joined the Masons as a Shriner.

Hill's wife became a tireless volunteer with the Imperial Order of the Daughters of the Empire during their time in Victoria, and for years held the position of president of the Esquimalt chapter until her death, on October 26, 1932.

Hill's health seemed to wind down in concert with the financial health of the Hillcrest mine. In November 1940, after the mine had closed, the ill seventy-eight-year-old widower travelled to California, where he expected to spend the winter in a climate more forgiving of old bones. Hill had just reached the home where he was expected as a guest when he collapsed and died. His heart had given out. Hill's remains were sent back to Victoria for burial next to his wife. Hillhaven was demolished in 1971, but the cove on which it sat remains much the same as it did in Hill's day. Anyone who stands there can look out toward the Juan De Fuca Strait just as Hill did for nearly twenty years.

Hill had forged the Hillcrest mine into the foundation of his life and financial success. He had found the highest quality coal in the pass, found investment to begin the mine, built it from nothing into a viable business, and sold a controlling share, while retaining enough shares to earn himself a substantial ongoing income. When it dissolved into history on April 10, 1940, the sick old man must have realized that his work was done. Whatever people thought of him personally, he had made the Hillcrest property into a valuable business that provided a living for thousands of men and their families, supplied fuel for locomotives, steam shovels, rollers, and farm tractors, heated homes and other buildings for the thirty-five years it was needed most, and created wealth for himself and others. It was a legacy that any man could look to with pride. That 189 men had died in the mine after he gave up control must have sat quietly in the back of his mind, always with him like a shadow on a bright day.

Perhaps the only still-functioning piece of equipment used by Hillcrest Collieries at the time of the explosion is the dinky steam locomotive that hauled loaded coal cars from the mouths of the no. 1 and no. 2 slants to the tipple. The locomotive now pulls children and their parents through a historical theme park in Saskatchewan, its role at the scene of Canada's worst mine disaster largely forgotten.

The ruins of Hillcrest Collieries are still there on the ridge above the town. They remain on private property, and visitors are discouraged with a "No Trespassing" sign. The walls of the wash house have fallen in, except for one, which leans against a tree for support. The floor on which the feet of countless miners walked and showered still lies smooth to the touch.

Nearby sit the walls of the engine and power house, where steam and electrical power were generated for the town, mine pumps, lights, and the no. 1 fan and hoist. The roof is long gone, the interior open to the sky, and trees grow throughout as the forest reclaims the place. What was once a cauldron of noise and heat, where steam hissed and coal fires glowed hot and red, fed by muscle and sweat, and where ashes from the fire boxes were dumped into the swiftly flowing water in the gutters that passed in front of the boilers, now bushes grow and birds nest and twitter. Just above, up the slope, the shell of the no. 1 hoist house still stands, and within, the empty pedestals for the missing hoist. Someone has fired a high-powered rifle into the north wall, and light filters through a gaping hole in the rotting concrete. In front, on the side that used to face the mine, water still seeps out of the earth from the workings, a few wooden rail ties lie covered in dead vegetation, and a moss-covered and rusting coil of wire rope sits forgotten near the old entry into the rock tunnel. If you walk between the engine house and the hoist house along the rail bed, to where the no. 2 entry plunged into the hillside beside Drumm Creek, you will find that water there, too, seeps from the ground from the buried workings, and stinks of sulphur. Up the hill, sinkholes extend into the forest where the workings have collapsed. In 1914, Drumm Creek flowed through a pipe under the no. 2 mine yard, but now the creek flows uncovered, and the mining detritus protrudes from the banks: rusting cable, pipe, and unidentifiable pieces of metal. The remains of a concrete and wooden trestle crumble into the creek. On the other side of the creek, only the foundations of the surface buildings remain. A little farther up the hill sits a solid, squat, thick-walled concrete structure not much larger than a garden shed. This was where blasting materials were stored.

Deer and elk now roam the forest, which has reoccupied the slopes above the mine. But if, on a clear day, you stand on the trail that curves up from the no. 1 mine yard past where the no. 1 fan sat, you may hear, just for a moment, what sounds like machinery in the mine yard. But the sound is just as quickly carried away on the breeze, and you will hear again only the sounds of the forest and your own breath.

GLOSSARY OF COAL MINING TERMINOLOGY

Air box	Wooden tubes that conveyed fresh air from the fan into the mine workings.
Blackdamp	A mixture of unbreathable gasses left after oxygen is removed from the air that typically consists of nitrogen, carbon dioxide and water vapour.
Brattice	A coarsely woven cloth saturated with tar, hung in passageways, that was used to control air flow and remove methane or firedamp.
Brattice man	The worker who installed the brattice in a mine.
Bucker	The worker who kicked, pried, or shoveled coal blockages in the chutes in an effort to dislodge them.
Bump	A sudden jarring of the mine produced by the giving way or cracking of the strata above or below the coal seam.
Check (or mine check or lamp check)	A numbered brass disc about the diameter of a one-dollar coin.
Chute	A wood-framed, sheet steel-surfaced slide into which coal is shoveled by the miners, built from the face where coal is extracted in a room, down to the level,
Coke	A fuel with few impurities, derived from low-sulfur bituminous coal.
Company man	A regular salaried employee of the colliery and not a member of the union, as opposed to a contract miner who was paid only on the basis of how much coal he mined. The company, non-union men included pit bosses, fire bosses, the boss driver, the stable boss, the master mechanic, electricians, weighmen, head carpenters, head blacksmiths, foremen, timekeepers, coal inspectors, and the head lampmen.
Counter (counter slope)	At Hillcrest Collieries, this referred to the new slope driven uphill to the rock tunnel. It would eventually come to be called the no. 1 slant.
Creep	The forcing up of the floor in the mine road and workings by the pressure of surrounding beds.

Dinky	The small steam locomotive on the surface that hauled mine cars full of coal to the tipple, and pulled the empty cars back to the mine entries.
District	The area of a mine defined by its ventilation.
Downcast	The passage through which the downward current of fresh air flowed into a mine.
Drift	Any horizontal passage underground.
Driver	The man who drove the horses that pulled full cars or trams of coal to the slant for haulage to the surface, and took the empty cars back to the chutes.
Driver boss	The man responsible for underground drivers, buckers, and car loaders.
Engine house (power house)	The large buildings on the surface that contained the steam boilers, which powered the electrical generators, and at no. 2 slant supplied steam to power the fan and hoist.
Face	The exposed surface of a coal seam from which the coal was extracted.
Fan boy	The boy who operated the hand fan used to expel firedamp or methane from areas of the mine not cleared by the usual ventilation techniques.
Fire boss	A provincially certified supervisory mine official employed by the mine company to examine the mine for combustible gases and other dangers before a shift of workers entered into it. Also responsible for handling and the ignition of Monobel in shot firing.
Heading	A main road in or out of a district in the mine.
Hoist	The powered cable drum that pulled mine cars full of coal up the slant to the surface and lowered the empty cars back into the mine. At Hillcrest Collieries, the No. 1 hoist was electrically driven, while the No. 2 hoist was steam powered.
Hoist house	The buildings that housed the hoists.
Level	A mine tunnel or gangway with track on the floor for the mine cars. Levels were driven off the slants with a shallow grade up to the rooms, allowing horses to easily pull empty mine cars from the slant to the bottom of the chutes in the rooms, and mine cars full of coal back down the level

to the slant. There, the cable from the hoist was attached to a string of laden cars and hoisted up the slant to the surface.

Monobel	The explosive used in Crowsnest Pass to blast coal loose at the face. It was considered safer than dynamite because it produced no flame, and was therefore considered less likely to ignite methane or coal dust.
Outbye	Towards the main road in or out of a district.
Outcrop	The part of a coal deposit exposed at the surface.
Overman (pit boss)	The third-highest ranking officer of the mine. Overmen had the constant charge of everything underground, examined the ventilation, and kept an account of all proceedings underground.
Parting	A junction in a roadway or tram road.
Pillar	A column of coal left to support the overlying strata in a mine, generally resulting in a room and pillar array. Coal pillars, such as those in pillar and stall mining, were extracted later, allowing the roof to collapse in a planned fashion.
Pillar and stall	A system of mining coal in which solid blocks of coal are left on either side of a miner's working place to support the roof until first mining has been completed, when the pillar coal is then recovered. Sometimes called "room and pillar."
Pitch	The angle at which the coal measure slanted into the earth.
Pump man	The worker who installed and serviced the water pumps used to pump water from the mine.
Raise	An inclined, vertical, or near-vertical opening driven up-ward from one level to connect with the level above, or to explore the ground for a limited distance above one level.
Return airway	A roadway through which stale air passed to the upcast.
Rib	The side of a pillar, the wall of an entry, or the solid coal on the side of any underground passage.
Rob	To extract pillars of coal previously left for support.
Room	A tunnel driven up the pitch from the slant, with a chute from top to bottom. The coal mined at the face at the top of the room, which had the effect of driving the room up

the coal seam, was shoveled into the chute, and slid down the length of the room to the mine car that sat at the bottom of the chute on the level.

Room and pillar See pillar and stall.

Rope rider The worker at the top of the slant who coupled and uncoupled mine cars from the wire rope attached to the hoist.

Slant (slope) The tunnel into the mine that slanted downward, following the coal seam. It served as the entry and exit point for workers, and, with mine car tracks on the floor, was where coal was hauled to the surface. Electrical cables for lights and pumps were also laid in the slant, as were the water pipes used to carry out water pumped from the mine.

Split Any division or branch of the ventilating current, or the workings ventilated by one such division or branch.

Sprag A short, pointed length of wood inserted into wheels to prevent cars from moving on an incline.

Timberman The worker responsible for setting props in the slants and levels.

Timber packer The worker responsible for bringing the wood for props to the workings.

Tipple The large structure common to all coal mines. At Hillcrest, laden coal cars were hauled to the tipple from the mine entries by the dinky. A mechanical dump emptied the cars, and a conveyor lowered the coal to the screens, where it was separated according to size, then loaded into closed CPR cars for shipment.

Tracklayer The worker who laid steel track in the mine for mine cars, and who removed track when an area was mined out or disused.

Trip A train of mine cars coupled together for movement along the level, or up or down the slant. A man-trip carried men into or out of the mine.

Upcast The passage through which stale air is returned to the surface.

TIMELINE OF EVENTS

1905

The Hill Crest Coal and Coke Company, Mine #0040, begins operations.

NOVEMBER 3, 1905

Tunnels have been driven in two seams. Shipments will begin as soon as the short rail line, now under construction from the CPR tracks, reaches the mine.

MARCH 22, 1906

The railway is now completed. About 16 men are employed belowground.

JUNE 7, 1906

The #1 seam produces 150 T. per day, employing 40 men in two shifts. The ventilation is by a furnace at the bottom of the upcast.

SEPTEMBER 6, 1906

The Deputy Minister orders the general manager of the Hillcrest Mine, T. MacLane, to discontinue the use of the furnace for ventilation at once, and immediately ventilate the mine with a fan. He notes that miners have complained about the ventilation.

SEPTEMBER 19, 1906

District Mines Inspector Elijah Heathcote reports that the furnace is discontinued, and natural ventilation will be used until a fan is installed. The District Inspector of Mines gives the mine authority to employ thirty-seven men until the fan is in place.

NOVEMBER 15, 1907

Open lights are used throughout, and the output of coal per day is 500 tons.

1907

Operations are suspended at the mine during the summer due to a strike by coal miners in B.C. and Alberta. Some development work continues at the mine during the strike which ends on July 24. Plans for a town site are prepared. Coal sales for the year reached $152,000. Wages paid out for the year amount to $91,046.00.

Hill publishes a brochure to promote the mine and town.

1908

Frederick Matthew Pinkney of Durham, England is appointed coroner for the area from Pincher Creek to Coleman. Previous to his appointment, he loses one hand in a zinc smelter accident. Until the accident, he is an accomplished musician on organ and violin.

FEBRUARY 7, 1908

Elijah Heathcote, reports to Deputy Minister John Stocks, that he has spoken with C.P. Hill about the installation of safety lamps. Hill says he will order them from Lethbridge.

FEBRUARY 10, 1908

Two miners are severely burned by an explosion of gas in the Hillcrest mines, ignited by open lights. Hill fails to report the accident as required by law. The accident prompts the order to use safety lamps exclusively in the mines.

APRIL 6, 1908

Hillcrest miners argue that there is very little gas in the mine, and that the use of the safety lamps "would be a considerable reduction to us under the existing agreement." A union official admits the danger from open lights, and two other officials admit that monetary considerations are the only reason against using safety lamps.

MAY 22, 1908

Elijah Heathcote reports that safety lamps are used exclusively.

SEPTEMBER 3, 1908

A fire destroys the lamp house and the safety lamps. C.P. Hill asks via telegram if he may work his mine with the open lights. Heathcote denies permission.

SEPTEMBER 8, 1908

C.P. Hill writes to the provincial government that "Mr. Heathcote should be taught that his duty is not to harass this Company any further." Hill is angry that in the absence of safety lamps Heathcote closed the mine. Hill says the lamps are not necessary, and even the union did not want them.

SEPTEMBER 28, 1908

Heathcote informs the provincial government that miners are not willing to pay to use lockers in a wash house that Hill proposes to build. He says Hill has been vague about how long it would be before a new wash house is built. Heathcote writes that Hill told him that he would to try to get him fired for his refusal to allow Hill to use open lights in the mine after fire destroyed the lamps.

SEPTEMBER 30, 1908

Again an angry Hill writes to the provincial government that there is no need for safety lamps, and that he has submitted to every request of the department at great expense to the Company which has never yet paid a dividend. Hill writes, "I am now tired of listening to the second hand complaints of our socialist mine committee as presented by your Mr. Heathcote." Hill charges that Heathcote was drunk in the Imperial Hotel in Frank on September 18, and said to the pit committee that he would compel Hill to settle the case as they wished it.

NOVEMBER 23, 1908

The union asks for help to force Hill to build a wash house.

DECEMBER 1, 1908

Stocks replies to the union secretary stating that Hill has promised to have a wash house built to comply with the law, and that the Inspector of Mines will visit the mine shortly to urge Mr. Hill to proceed on the construction as soon as possible.

DECEMBER 19, 1908

Hill writes a letter to the head of the CPR's Mining and Metallurgical Department, and managing director of the CPR smelter at Trail, W.H. Aldridge, and tells him that he intends to sell Hill Crest Coal and Coke for health reasons, and as promised, Aldridge is given the first option to buy.

1909

An eight-hour working day becomes law in Alberta.

MARCH 31, 1909

Workers strike throughout the District over a new contract.

JULY 19, 1909

The strike ends and mining resumes at Hillcrest Collieries.

DECEMBER 2, 1909

Hill writes to John Stocks, the Deputy Minister, that as of Dec. 1, 1909, James Sommerville Quigley will manage the mine.

DECEMBER 3, 1909

Aldridge writes to John Brown at the Bolen-Darnall Coal Co. in Hartford, Arkansas that friends have an option on control of Hillcrest. He offers Brown management at an annual salary of $5,000 if the "friends" exercise the option, and asks when is the earliest that he could reach Hillcrest upon acceptance of the offer.

APRIL 6, 1910

Hill moves out of his residence near the mine and leaves to tour Europe with his family, but remains a principle shareholder. Slope number 2 is begun in the summer.

APRIL 8, 1910

John Brown arrives at Hillcrest and takes over as General Manager of the mine.

MAY 10, 1910

William Hutchison arrives at Hillcrest.

DECEMBER 9, 1910

An explosion at the Bellevue Mine kills 30 men.

JANUARY 28, 1911

Elijah Heathcote reports that the wash house is in use.

1911

A strike interrupts work for nine months, but during the work stoppage a brick lamp house, an office for the fire bosses and mine manager, blacksmith and carpenter shops, and car barns are erected and equipped.

1912

A new steel tipple is completed by a firm of American contractors.

The first course of training in mine rescue in the Crowsnest Pass is held.

JULY 24–25, 1912

District Inspector of Mines Francis Aspinall inspects Mine #0040 (Hillcrest Collieries) and reports, "The ventilation in this mine is in my opinion inadequate, as I found a number of places with dangerous quantities of gas present." A new fan has been shipped and will soon be installed.

DECEMBER 11, 1912

The excavation for the cellar of the Cruickshank/Burnett store and Masonic hall in Hillcrest is almost complete. The Coal company installs electric lights for the town.

APRIL 16, 1913

The new Lodge hall, a dining room, reception room, and dressing room are opened. The Hall, with piano, will be used for social occasions as well as for Lodge work.

OCTOBER 1, 1913

Francis Aspinall leaves the Crowsnest Pass as District Inspector of Mines.

JANUARY 31, 1914

The new Crowsnest District Inspector of Mines, Andrew Scott, inspects Mine #0040. He finds the general condition of the mine good, and the condition of ventilation good.

The population of the town is about 1,000.

APRIL 6, 1914

Scott visits Mine #0040. In the No. 2 mine he finds the ventilation and timbering good and no explosive gas.

MAY 18, 1914

The pit committee, made up of three workmen appointed or elected by the United Mine Workers of America finds gas in 41 & 42 rooms south, but ventilation good, and general conditions good.

JUNE 17–18, 1914

The Hillcrest mine is closed for the two days because demand is down, although inspections continue.

JUNE 18, 1914

Fire-boss Dan Briscoe comes out of the mine at 10:30 PM. He claims that he finds *only a small quantity of gas in No. 3 south entry.* "Ventilation was as usual. Coal dust was normal and lots of moisture in the mine. This mine is run in a careful manner as to safety and cannot be called a dusty mine," he testifies at the inquest.

JUNE 19, 1914

7 AM W.P. Adlam, fire-boss, goes off shift. During his inspection he finds two caves. He claims to find indications of gas but no more than usual. The barometer indicates a larger volume of gas in mine. He posts his report as usual. Later he says he found enough gas to put his light out in six rooms which he fences off, and reports.

Fire-bosses John Ironmonger and Sam Charleton inspect the lamps of the miners and find them in good shape, then Ironmonger enters the mine with the Superintendent J.S. Quigley. They become separated. Ironmonger examines the mine, but finds no traces of gas or caves.

228 men enter the mine, including the boys who operate the hand fans for ventilating the raises.

8–9 AM Ironmonger fires five shots to loosen coal.

9 AM Seven more men enter the mine through the rock tunnel. There are now 235 men in the mine. Two other men who take checks do not enter the mine, but work outside in the mine yard.

9:15–9:30 AM The Hillcrest mine explodes. Ironmonger testifies at the inquiry: "The first intimation I had of anything wrong was a concussion. There was no noise. I started to go out and met smoke coming in the entry. There was no heat."

Miner Yuriy (George) Frolick had delivered eight carts of coal to the No. 2 slant, and is getting his horse to haul a train of eight more carts to the slant where the hoist will pull the carts to the surface. The concussion from the explosion knocks Frolick unconscious.

Chief Engineer William Hutchinson is on the No. 3 slope, an abandoned part of the mine, when the explosion occurs. He and his brother David are enveloped in a tremendous wind in which they can hardly stand up. The wind is followed about thirty seconds later by brown smoke. He hears no sound of an explosion. He says he is out within a minute near the mouth of No. 2 slant which is belching smoke. The explosion has stopped, but not damaged the fan. Hutchinson and the hoist engineer, Tom Brown, reverse the doors and start the fan again.

Hutchinson informs Brown that he will enter No. 2, and plunges into the mine.

The blast demolishes part of the eight-inch-thick concrete wall of the No. 1 engine house which faces the mine, and blows its roof off. Nineteen-year-old Charles Ironmonger is blown from the mouth of No. 1. He dies after admission to hospital.

The mine's steam whistle blasts, announcing the emergency in the mine.

9:45 AM Frolick regains consciousness. He finds it difficult to breathe and the heat intense. The lamps have been extinguished by the blast, leaving the miners in darkness. Frolick grabs the mane of his horse, and half walks, half lets himself be dragged along, trusting the instinct of the horse to know the way out, and shouts at others to follow.

10 AM Half an hour after the explosion, the first organized rescue party without oxygen descends into the mine.

Henry James, the superintendent of Mine Rescue Car No. 1, located at Blairmore three miles away, receives word by telephone from Hillcrest Mine for the rescue car to be rushed to the Hillcrest Mine at once.

Crowds of anxious relatives and friends of the miners begin to congregate at the mines.

The No. 1 fan is restarted.

10:15 AM The mine rescue car arrives at Hillcrest.

10:30 AM Inspector Christen Junget, Pincher Creek commander, receives a telephone call that a serious explosion has occurred in the Hillcrest mine and that assistance is needed. He telephones Superintendent Cortlandt Starnes, commander of "D" Division at MacLeod, to inform him, and phones the RNWMP at Lundbreck, Bellevue, Coleman, and Blairmore. Junget then leaves Pincher Creek immediately by motor car with Corporal Searle and Constable Kistruck. Upon his arrival, Junget discovers that Corporal Mead has taken the first steps in handling the crowd, and retrieving the dead from the mine.

Women in surrounding communities rush to assist the rescue efforts and the relief of families affected by the disaster. Women from Bellevue and other points are on the scene within an hour with refreshments for the rescuers: galvanized buckets of coffee and tea.

12:30 PM The mine is practically clear of noxious gases.

12:55 PM The special train supplied by the C.P.R. departs Lethbridge carrying eight mine rescue teams from Lethbridge and Coalhurst.

2:01 PM A CPR train arrives at Hillcrest having departed Fernie at 12:10. The train brings medical personnel and the B.C. mine rescue car, which carries the B.C. government's Draeger rescue apparatus from the Fernie station, all the available apparatus of the Crow's Nest Pass Coal Company at Coal Creek, and seven trained men from Hosmer.

2:30 PM Ten dead bodies are brought up from the mine on lumber trucks covered in blankets, and are transferred to coal cars.

3 PM Hope is abandoned that anyone remains alive in the mine. In all, the Pulmotor was tried on sixteen men who were brought out of the mine. It saved the lives of eight of them.

3–4 PM The train from Lethbridge arrives with Stirling and the mine rescue teams from Lethbridge and Coalhurst. Also on the train are six nurses from Lethbridge, and members of the St. John's Ambulance Corps. About 20 bodies are now recovered.

4 PM Top pump No. 2 is restarted.

Bellevue miner Harry White finds fires. The loose coal on the pavement is burning.

8 PM Fifty-two bodies have been recovered. Work continues into the night.

11:30 PM The rescue gangs all are recalled after one party encounters more fires. The rescue party from Hosmer, led by Mr. Shaw, is sent down to investigate the conditions before any more rescue parties are allowed to go down. A call for more men and water is sent to the surface.

At about **3 AM** another call for volunteers is sent up. Only the already exhausted Lethbridge crew comes forward: J. Stevenson, Ed Berford, Aleck Stevenson, A. Quinn, G. Coutts, W. Goldie, G. Hargreaves, Louis Moore and district inspector Sam Jones.

JUNE 20, 1914

The small fires burning in the mine prove difficult to extinguish. John Stirling counts 14.

When RNWMP Superintendent Starnes arrives at Hillcrest early Saturday morning, he is met by Inspector Christen Junget who suggests that the bar of the local hotel be closed. Starnes wires the Deputy Attorney General. The hotel keeper is told that his licence is suspended until Monday night. The suspension would later be extended until June 26.

It is payday at the mine, but the company issues no pay.

The Coroner's Inquest meets with a jury of ten men headed by Coroner Frederick Matthew Pinkney, views bodies and concludes they were killed as a result of an explosion of gas and coal dust. Up until 10 AM a total of eighty-six dead bodies have been taken from the mine.

In the morning, two carloads of coffins arrive from Calgary and MacLeod.

A crew of some 40 men is put to work digging graves.

JUNE 21, 1914

In the morning, at the Miners' hall, bodies are placed in coffins, and carried out to the common where they are labelled and placed side by side in rows.

Services are held by clergymen of all denominations with the burial of most of the bodies in the afternoon.

117 men are buried this day at Hillcrest.

In the afternoon, all of the fires in the mine are finally extinguished.

JUNE 22, 1914

All but eight of the 189 bodies in the mine have been recovered. The survivors are listed as 48.

Thirty-two more bodies are buried at Hillcrest. A severe windstorm creates havoc in the Pass. Telegraph and electrical wires are broken, and the roofs blown off buildings.

The MLA for Rocky Mountain, Robert E. Campbell, appeals to Canadians for assistance to alleviate the distress of the victims' families. He appoints a relief committee.

JUNE 23, 1914

Twelve bodies are buried at Hillcrest.

The City of Lethbridge donates and dispatches to Hillcrest two rail car loads of provisions.

The Empress movie theatre in Calgary and the Starland in Lethbridge both screen slides taken at the disaster scene.

JUNE 24, 1914

Three more bodies are buried at Hillcrest.

An Order in Council of the Alberta Government appoints Judge Arthur Allan Carpenter to act as Commissioner to conduct an investigation into the cause and effect of the Hillcrest explosion.

A relief store, stocked in large part by a rail car of supplies donated by the city of Lethbridge, opens in Hillcrest.

The president of the CPR, Thomas Shaughnessy, wires P.L. Naismith in Calgary that the CPR will contribute $2,000 toward the relief of families of men who lost their lives in the Hillcrest disaster.

JUNE 26, 1914

A list of the names of 186 recovered bodies is released. The three remaining bodies are said to be Oakley, Bainbridge, and an unknown.

John Stirling writes to the Deputy Minister, John Stocks that for relief of families left destitute by the disaster, T. Eaton Co. has donated $1,000, CPR $2,000, the City of Calgary $2,500, and the City of Lethbridge $1,000 in addition to the supplies sent.

JULY 2, 1914

10 AM The Commission of Enquiry into Hillcrest Mine Disaster commences in the Masonic Hall above George Cruickshank's General Store. Evidence is given under oath.

JULY 3, 1914

The Province of Alberta, by Order in Council, appoints a permanent Commission called the Hillcrest Relief Commission, and grants it $20,000 for the relief of widows and orphans left dependent owing to the death of miners killed at Hillcrest on June 19.

JULY 4, 1914

10 AM The Commission of Inquiry resumes sitting.

3 PM The body of William Oakley is recovered. It is found buried under a pile of coal and rock. The body is sent to Michel, B.C. for burial.

JULY 7, 1914

The mine resumes operations for the first time since the explosion. 133 tons are produced on the first day. About 80 men work per shift.

JULY 11, 1914

The Commission of Inquiry under Judge Carpenter concludes its hearings.

JULY 21, 1914

10 AM The Coroner's Inquest opens at Masonic hall in Hillcrest before F.M. Pinkney, Coroner. The testimony of fourteen men is taken in shorthand over two days by Court Reporter T.A. Powell.

JULY 29, 1914

The RNWMP Commissioner requests that the conduct of Corporal Mead, Corporal Grant, and Constable Hancock be mentioned to the Prime Minister, and that they each be given a grant of $50 in recognition of their work.

JULY 31, 1914

Within six weeks of the disaster, 140 men arrive to take the jobs left vacant by the fatalities.

MID-AUGUST, 1914

Hillcrest miners strike because management has prohibited them from shot-firing.

FEBRUARY 23, 1915

Hillcrest management conducts a search of their underground workmen. Three men have tobacco or matches on them. The District Inspector of Mines recommends to the Chief Inspector of Mines that the men be prosecuted.

JUNE 1, 1915

Hillcrest Collieries Ltd. pays the first instalment of the compensation to the widows and orphans of the Hillcrest Mine Disaster. Full compensation is expected to be paid to more than ninety families.

JUNE 15, 1915

Hillcrest miners strike to exclude Austrian and German miners following the outbreak of WWI.

JUNE 19, 1915

Hillcrest Collieries reopens, but tensions escalate as 150 German and Austrian miners return to work.

MID-JULY, 1915

The *District Ledger*, the organ of the United Mine Workers Union, publishes a long article opposing the internment of German and Austrian miners as was done at Hillcrest and Fernie.

JANUARY 6, 1920

The Lethbridge Herald reports that the United Mine Workers of America has found irregularities in the books of District 18 which was under control of members of the One Big Union (OBU). *The Herald* reports that $548.30 from the Hillcrest Relief Fund went missing into personal accounts of the former District 18 OBU leadership, and was only returned to the Fund after a letter was written by the UMWA international commission. Another $1,870 of the Hillcrest Relief Fund was transferred to the General Fund and never paid back.

1921

C.P. Hill and his wife Enid move from Montreal to Victoria, B.C.

SEPTEMBER 19, 1926

A massive explosion, even worse than the 1914 blast, rocks the Hillcrest Collieries mine, but only two men are inside. Fifty-five-year-old fire boss Frank Lote and twenty-five-year-old pump mechanic Fred Jones are killed. The cause is determined to be a spark set off by a fall of rock which ignited a gas explosion, and subsequent coal dust explosions.

OCTOBER 26, 1932

Enid Hill dies at 7 AM at the family residence "Hillhaven" on Esquimault Road in Victoria, B.C. She has been in poor health for some time, but her death is "unexpected."

DECEMBER 4, 1939

Notice is given of intention to abandon the Hillcrest mine within the next ninety days.

APRIL 10, 1940

The Hillcrest mine is abandoned, effective March 31, 1940. All entrances to the underground workings are sealed in accordance with the requirements of Section 100 of the mines Act.

NOVEMBER 29, 1940

C.P. Hill dies at Pasadena, California at the age of seventy-eight. His body is shipped to Victoria, B.C. for burial at Royal Oak Cemetery next to his wife.

KEY PERSONNEL AT HILLCREST COLLIERIES, CIRCA 1914

General manager: John Brown

Mine manager: James Sommerville Quigley (killed)

Overman (pit boss): Thomas Taylor (killed)

Examiners (fire bosses): William Adlam, Dan Briscoe, Sam Charleton (killed), John Ironmonger, Dan Kyle, Frank Lote, Walter Rose

Chief engineer: William Hutchison

Surveyor: David Hutchison

Brattice men: David Harris (killed), William Twohey, David Emery (killed), James Petrie (killed), Alex Petrie (killed)

Timekeeper: Robert Hood

Pump man: Thomas Brown (killed)

No. 1 hoist engineer: Unknown

No. 2 hoist engineer: Thomas Brown

Dinky engineers: George Porteus, Torrie Hood

Master mechanic: Thomas Hargreaves

Power house engineer: Ed Keith

Electrician: Andrew Wilson

NAMES OF MEN KILLED

The spellings of names are taken from government and company records created in 1914. Many names will vary from current spellings.

Peter Ackers
Herbert Adlam
Dominic Albenese
Nicholas Albenese
Robert Anderson
Jacob Andreaschuk
George Androski
James Armstrong
Sidney Bainbridge
Andrew Banlant
Steve Banyar
James Barber
Thomas Bardsley
Fred C. Bennett
Fred Bingham
Virgilio Bodio
John Bolinski
Frank M. Bostock
Etalleredo Botter
John S. Bowie
Pietro Bozzer
James Bradshaw
John Brown
Thomas W. Brown
William Brown
Albert Buckman
Joseph Camarda

Peter Cantalline
Antonio Carelli
Henry Carr
Carlo Cassagrande
Sam Cataline
Antonio Catanio
Basso Caterino
Vito Celli
Emil Chabillon
Leonce Chabillon
Charles S. Charles
Sam Charlton
Eugenio Ciccone
Antonio Cimetta
John Clarke
Leonard Clarke
Charles Coan
Thomas Corkill
Fred Coulter
Robert Coulter
Thompson Court
Dan Cullinen
John Davidson
Prosper Daye
George Demchuk
Nicholas Demchuk
Matthew Dickenson

Andrew Dugdale
Robert Dugdale
Charles Elick
David Emery
Everard Eveloir
James Ewing
Peter Fedoruk
William Fines
August Flourgere
John C. Fogale
Luigi Fortunato
Vincenzo Fortunato
John Foster
William Fox
Gustaf Francz
Frank Frech
William Gallimore
Emil Garine
Antonio Gasperion
Carlo Gianoli
Antonio Gramacci
James F. Gray
Ylio Guido
Ralph Hansford
David C. Harris
John Heber
Alphonse Heusdens

George Hicken

William Hillman

Phillip Hnacnuk

John Hood

Hugh Hunter

Wasyl Iluk

Charles Ironmonger

Sam Ironmonger

Mike Janego

Carl Johnson

Fred Johnson

William Johnson

Patrick Kane

Peter Kinock

Mike Kipryanchuk

Petro Kohar

Chris Kosmik

Dan Kostyniuk

Fred Kurigatz

Nick Kuzenko

Fred Kwasnico

Wasyl Kwasnico

Frank Labonne

Antoine Legard

Ulderico Marchetto

Cuiseppi Marcolli

Rod McIsaac

Angus H. McKay

John B. McKinnon

Steve McKinnon

Pius McNeil

John A. McQuarrie

Nicholas Megency

Adam Meiklejohn

Steve Melanchuk

John Melok

William Miller

William G. Miller

Dominic Montelli

William Moore

William Morley

Alex Morrison

Nick Morron

John Mudrik

Robert Muir

Fred Muirhouse

David Murray, Jr.

David Murray, Sr.

Robert Murray

William Murray

Steve Myrovich

William Neath

Joseph Oakley

Eduardo Pagnan

Arthur Pardgett

Carlo Parnisari

Cuiseppi Parnisari

Leon Payet

John Pearson

James Penn

Robert Penn

Alex Petrie

James Petrie

Robert Petrie

Alex Porteous

James Porteous

George Pounder

James S. Quigley

Thomas Quigley

Steve Raitko

Bernard Ralnyk

Fred Ralnyk

Albert Rees

George Robertson

Joseph H. Rochester

William Rochester

Eugenio Rossanese

Luigi Rosti

Alfred Salva

John Sands

John Sandul

Daniel Sandulik

Charles Schroeder

Mike Skurhan

Robert Smith

Thomas Smith

Peter Somotink

Albert Southwell

Edward Stretton

Albert Tamborini

Baldo Tamborini

Thomas Taylor

John Thaczuk

David O. Thomas

William Trump

Thomas Turner

William Turner

Mike Tyron

Fred Vendrasco

Joseph Vohradsky

Vince Vohradsky

David J. Walker

Rod Wallis

Thomas L. Wilson

John Zahara

Luis Zamis

Wasyl Zapisocki

Michael Zaska

NATIONALITIES OF MEN KILLED

Englishmen, Welshmen, Irishmen: 43

Italians: 35

Canadians: 27

Ruthenians (now called Ukrainians): 26

Scots: 26

Frenchmen: 7

Hungarians: 5

Americans: 4

Bohemians (immigrants from Bohemia in what is now the Czech Republic): 4

Slovaks: 4

Belgians: 2

Poles: 1

Swedes: 1

Others: 4

NOTES

CHAPTER 1: THE LURKING THREAT

1 Rice, George S. *Accidents from Falls of Roof and Coal*. Washington, D.C., Department of the Interior, Bureau of Mines, 1912, p. 5.

2 William Hutchison Fonds, M558, Glenbow Archives.

3 H.B. Humphrey. *Historical Summary of Coal Mine Explosions in the United States, 1810–1958*. Washington, U.S. Govt. Print. Off., 1960, p. 17.

4 Lamb, H. Mortimer, ed. *The Journal of the Canadian Mining Institute 1910*. Montreal. The Canadian Mining Institute, 1911, pp. 51–52.

5 Lamb, H. Mortimer, ed. *The Transactions of the Canadian Mining Institute 1913*. Montreal. The Canadian Mining Institute, 1914, pp. 53–57.

6 Humphrey, p. 32.

7 Bercuson, David Jay, ed. *Alberta's Coal Industry 1919*. Calgary, Historical Society of Alberta, 1978, p.156.

8 Humphrey, pp. 32–33.

9 Rice, p 4.

10 Humphrey, p. 42.

11 Cousins, William James. A *History of the Crowsnest Pass*, Edmonton, University of Alberta Press, 1952, p. 81.

12 Larry Ewasen oral history project. ca. 1980. M6833. Provincial Archives of Alberta.

13 Francis Aspinall and Andrew A. Millar, "Mine Rescue Apparatus, and the Value of Mine Rescue Work," in *The Transactions of the Canadian Mining Institute 1913*, ed. H. Mortimer Lamb (Montreal, The Council at the Secretary's Office, 1913) p. 463.

CHAPTER 2: LIFE AND IDEAS IN THE PASS

1 Mouat, Jeremy. "James Baker and William Fernie: The Politics of Development in the Crowsnest Coalfields," in *The Forgotten Side of the Border: British Columbia's Elk Valley and Crowsnest Pass*. ed. Wayne Norton and Naomi Miller. Kamloops: Plateau Press, 1998, pp.15–23

2 Coke, formed when coal is heated in a coke oven in the absence of air, supplies the heat as well as the gases required for the reduction of iron ore in a blast furnace.

3 Eagle, John A. *The Canadian Pacific Railway and the Development of Western Canada, 1896–1914*. Kingston, McGill-Queen's University Press, 1989. p. 245.

4 Cousins, p. 76.

5 Provincial Archives of Alberta, 69.210, #559, #775, Justice of the Peace duties and actions, 1894–1926.

6 *Crowsnest and Its People*. Coleman, Crowsnest Pass Historical Society, 1979.

7 Larry Ewasen oral history project. ca. 1980. M6833. Provincial Archives of Alberta.

8 Cashman, Tony. *Singing Wires: The Telephone in Alberta*. Edmonton: The Alberta Government Telephones Commission, 1972, p. 111.

9 Cashman, p. 111.

10 Much of the information in this section is in the paper presented by C. Allen Seager of Simon Fraser University at the B.C. Studies Conference in 1981, F.H. *Sherman of Fernie and the UMW of A 1902–1909: A Study in Western-Canadian Labour Radicalism*, an excellent source of information about Frank Sherman and radical unionism in the Pass.

11 Report of the Vice President, United Mine Workers of America, 8th Annual Convention, District 18, Lethbridge, 1911, p. 60, Glenbow Archives, M2239.

12 Before he became mayor of Lethbridge, Hardie had been a mine manager, and as such was considered a mining "expert." He had written a paper titled "The Galt Coal Field, Alberta" published in *The Journal of the Canadian Mining Institute* 1910.

13 Aspinall had been district mine inspector for the Crowsnest region, but had changed to the Edmonton district by the time of the Hillcrest Disaster. He testified at the Inquiry and the Inquest.

14 Precedents had been set in the United States over and over again by unions, including mine unions who sometimes used dynamite to settle grievances which resulted in massive property damage and loss of life. On October 1, 1910 the Los Angeles Times Building was completely destroyed by an explosion of dynamite and fire in which 21 lives were lost. Two brothers, both members of the Iron Workers union, were held responsible and sent to prison for life. On March 8, 1911 in Springfield, Illinois, two dynamite explosions destroyed a span of a viaduct, and damaged a coal company tipple. These were intimidation tactics by union members, and represent a tiny fraction of the union violence in the United States at the time.

15 Provincial Archives of Alberta. 66.166, Box 22, File 357.

16 The unpublished memoirs of Jim Hutchison, Edmonton, 2007.

17 Luciuk, Lubomir and Carynnyk, Marco, ed. *Between Two Worlds: The Memoirs of Stanley Frolick*. Toronto: The Multicultural History Society of Ontario, 1990, p. 36.

CHAPTER 3: C.P. HILL: AN AMERICAN IN THE COAL FIELDS

1 *Who's Who and Why 1919–1920*, p.1067.

2 *Crowsnest and Its People*, p. 330.

3 *Alberta Past and Present: Historical and Biographical, Volume III*. Chicago, Pioneer Historical Publishing Co., 1924, pp. 45–46.

4 Memo from Norman Fraser to John Stocks, October 7, 1908, PA 77.237, 40j, Box 8.

CHAPTER 4: HILL SELLS

1 *Alberta Past and Present: Historical and Biographical, Volume III*. Chicago, Pioneer Historical Publishing Co., 1924, pp 413–414.

2 Letter from W.D. Matthews to W.H. Aldridge, Nov. 25, 1909, CPR M2269, file 1956, Box 198, Glenbow Archives.

3 CPR M2269, Box 198, file 1956, Glenbow Archives.

4 M.P. Davis of Ottawa was the contractor for the masonry substructure on the Quebec bridge across the St. Lawrence River. Seventy-five workers died on August 29, 1907 when the steel cantilever super-structure collapsed into the river while under construction. Davis was in no way responsible for the accident. Responsibility was attributed to the design engineer and the consulting engineer.

CHAPTER 5: THE BELLEVUE DISASTER

1 Hutchison, William. William Hutchison Fonds, M558. Glenbow Archives. p. 28.

2 Virtually all of the details of the Bellevue Disaster are detailed in the lengthy official transcript of the coroner's inquest held in the Provincial Archives of Alberta in Edmonton, Accession # 67.172, Box 11, file #594.

3 John T. Stirling and Prof. John Cadman, D. Sc., *The Bellevue Explosions*, 1912. Acc. #1977.237, File 87a.

4 John Powell, Bellevue Superintendent, Acc. #1977.237, file 87g, Provincial Archives of Alberta.

5 *Transcript of the Coroner's Inquest into the Bellevue Explosion, 1910.*

CHAPTER 6: THE DRIFT TOWARDS DARKNESS

1 Letter from Lewis Stockett to W.B. Powell, President of District 18, UMW of A, Oct. 26, 1911. *Coal Association of Canada Fonds*, M2210, File 37. Glenbow Archives.

2 *Crowsnest and Its People*, Coleman, Crowsnest Pass Historical Society, 1979, p. 521.

3 Aspinall co-wrote a paper titled "Mine Rescue Apparatus and the Value of Mine Rescue Work" published in *The Transactions of the Canadian Mining Institute*, 1913.

4 Interview with George Cruickshank conducted by CBC Radio, Calgary.

5 *Hillcrest-Bellevue Early Days: Souvenir Booklet.* p. 29.

6 *Crowsnest and Its People.* p. 798.

7 *Crowsnest and Its People.* p. 584.

CHAPTER 7: WITHOUT AIR TO BREATHE

1 Stanley W. Frolick as told to him by his father, Yuriy (George) Frolick. *Review, the Ukrainian Canadian Professional and Business Federation Magazine.* 1974. pp. 31–34.

2 By 1919 at the latest, the U.S. Bureau of Mines recommended that frogs be blocked with wood to prevent boots from getting caught. See: Edward Steidle, *Dangerous and Safe Practices in Bituminous Coal Mines.* 1919, Washington, U.S. Bureau of Mines, p. 36

3 *Edmonton Bulletin*, June 22, 1914, p. 2.

4 *The Calgary Daily Herald*, story written June 21, published June 22, 1914.

5 The unpublished memoirs of William Hutchison, William Hutchison Fonds, M558. Calgary, Glenbow Archives, 1959.

6 *Crowsnest and Its People*, p. 696.

7 *The Lethbridge Herald.* June 20, 1964.

8 Francis Aspinall and Andrew A. Millar, "Mine Rescue Apparatus, and the Value of Mine Rescue Work" in *The Transaction of the Canadian Mining Institute 1913.* Montreal, The Council at the Secretary's Office, 1913, p. 464.

9 *Alberta Past and Present: Historical and Biographical Volume II*, Chicago, Pioneer Historical Publishing Co., 1924, pp. 479–480.

10 Kelly, Nora. *The Men of the Mounted.* Toronto: J.M. Dent & Sons, 1949.

11 Transcript of the commission of inquiry, p. 43.

12 Details of Cruickshank's involvement are from an interview with him conducted for CBC Radio, Calgary by Delores MacFarlane, 1958–1962. Exact date unknown. Library and Archives Canada, Ottawa.

13 Transcribed from an audiotaped interview with George Loxton, Sir Alexander Galt Archives, Lethbridge.

CHAPTER 8: CRUICKSHANK'S SHROUDS

1 Interview conducted with Cruickshank for CBC Radio, Calgary.

2 Transcribed from an audiotaped interview with George Loxton, Sir Alexander Galt Museum and Archives, Lethbridge.

3 *Bellevue Times*, Nov. 21, 1913.

CHAPTER 9: THE REPORTERS

1 Anderson, Frank and Turnbull, Elsie G. *Tragedies of the Crowsnest Pass*. Surrey, BC, Heritage House Publishing Company Ltd., 1983.

2 *Lethbridge Herald*, June 22, City and District section.

3 *Edmonton Bulletin*, June 22, 1914, p. 2.

CHAPTER 10: THE DEAD AND THE BROKEN

1 *Crowsnest and its People*. Coleman, Crowsnest Pass Historical Society, 1979.

2 Buchanan was elected as a Liberal in the second Alberta general election of 1909, thanks largely, as he recalled later, to strong support among the ranks of the mine workers on the north side of Lethbridge. He was elected to a federal seat in 1911.

3 The Cherry Mine Disaster, November 13, 1909, ninety miles southwest of Chicago, Illinois. A fire began underground in a car filled with hay, and rapidly spread. 268 men died of burns and suffocation including twelve rescuers. twenty men walled themselves into a mine chamber and were rescued on November 20, despite the fact that the mine had been sealed to starve the fire of oxygen. For a detailed, though florid account of the Cherry Disaster, see *The Cherry Mine Disaster*. F.P. Buck. Chicago, M.A. Donohue & Co., 1910.

CHAPTER 11: Q&A

1 There is some discrepancy here: at the time of the Bellevue Disaster in 1910, White was a fire boss at Frank.

CHAPTER 15: CONCLUSIONS

1 Carpenter, A.A. *Report of the Commission Appointed for the Investigation and Enquiry Into the Cause and Effect of the Hillcrest Mine Disaster*. Edmonton: Government Printer, 1914, pp. 11–12.

2 Campbell, J.D., *Catalogue of Coal Mines of the Alberta Plains*. Edmonton, Alberta: Research Council of Alberta, Geological Division, 1964.

3 Results of an experiment by the author.

CHAPTER 16: ECHOES

1 Personal interview with Barbara Allen, 2007.

2 Personal interview with Barbara Allen, 2007.

3 Luciuk, Lubomyr Y. and Carynnyk, Marco, ed. *Between Two Worlds: The Memoirs of Stanley Frolick*. Toronto: The Multicultural History Society of Ontario, 1990.

4 Norton, Wayne and Naomi Miller, ed. *The Forgotten Side of the Border*. Kamloops: Plateau Press, p. 72.

5 Unpublished memoirs of Jim Hutchison, Edmonton: 2007

6 *Crowsnest and Its People*. Coleman: Crowsnest Pass Historical Society, 1979.

7 *The Lethbridge Herald*, January 6, 1920.

8 Blue, B.A., John. *Alberta Past and Present: Historical and Biographical*. Vol. 1. Chicago: 1924.

9 There is a bitter irony to the acceptance of Communist radicalism to fight the low pay during the Depression when the Socialist, or "progressive," policies of interference in the free market, particularly in the United States, in the form of FDR's New Deal, were directly responsible for prolonging the Depression. See Powell, Jim. *FDR's Folly: How Roosevelt and His New Deal Prolonged the Great Depression*, New York: Three Rivers Press, 2003. FDR tripled taxes, did nothing about the Smoot-Hawley tariff which choked trade, weakened the banking system with second banking "reform" the Glass-Steagall Act, and generally hindered economic recovery by intervening in labour markets to raise wages and labour costs. Another valuable resource that reinforces this position with recent scholarship is *The Forgotten Man: A New History of the Great Depression*. Shlaes, Amity. New York, Harper, 2007.

10 From 1874, when the police service first arrived on the prairies, to 1904, it was known as the North-West Mounted Police. In 1904, it became the Royal Northwest Mounted Police, and in 1920, the Royal Canadian Mounted Police.

CHAPTER 17: THE CLOSURE

1 *The Memoirs of Frank Hosek*, unpublished.

2 Personal conversation with Judge Carpenter's daughter, Mrs. Alice Jamieson of Edmonton.

3 Born in Orillia, Ontario in 1848, Sam Steele rode in the NWMP's 1874 march west, met with Sitting Bull, fought Big Bear at Loon Lake, brought order and law to the Klondike gold rush in the Yukon where he was Cortlandt Starnes' superior officer, commanded the Lord Strathcona Horse in the second Boer war, was knighted in Britain during WWI, died during the 1918 flu pandemic, and was buried in Winnipeg.

4 Fetherstonehaugh, R.C. *The Royal Canadian Mounted Police.* New York: Carrick & Evans, Inc., 1938.

5 The RCMP opposed the internment of Japanese Canadians during the war, and considered it unnecessary. The internments were, in fact, a political decision which pandered to public fears. The Liberal government under Mackenzie King gave into pressure from a few politicians in B.C. whose motives were base: to gain status and to ingratiate themselves to Ottawa. Mead at the time was the RCMP's security expert who spoke quite openly in favour of the Japanese Canadians. For an insightful account of the matter, see *The Politics of Racism: The Uprooting of Japanese Canadians During the Second World War* by Ann Gomer Sunahara, Toronto, James Lorimer & Company, Publishers, 1981.

6 *The RCMP Quarterly.* Volume 35, No. 1. Ottawa: July, 1969

BIBLIOGRAPHY

Bercuson, David Jay, ed. *Alberta's Coal Industry 1919.* Calgary: Historical Society of Alberta, 1978.

Bowen, Lynne. *Boss Whistle: The Coal Miners of Vancouver Island Remember.* Revised edition. Nanaimo: Nanaimo and District Museum Society and Rocky Point Books, 2002.

Cashman, Tony. *Singing Wires: The Telephone in Alberta.* Edmonton: Alberta Government Telephones Commission, 1972.

Cousins, William James. *A History of the Crowsnest Pass.* Edmonton: University of Alberta Press, 1952.

Eagle, John A. *The Canadian Pacific Railway and the Development of Western Canada, 1896–1914.* Kingston: McGill-Queen's University Press, 1989.

Fetherstonehaugh, R.C. *The Royal Canadian Mounted Police.* New York: Carrick & Evans, Inc., 1938.

Freese, Barbara. *Coal: A Human History.* Cambridge, MA: Perseus, 2003.

Hayek, F.A., ed. *Capitalism and the Historians.* Chicago: University of Chicago Press, 1954.

Humphrey, H.B. *Historical Summary of Coal-Mine Explosions in the United States, 1810–1958.* Washington: Bureau of Mines, U.S. Department of the Interior, 1960.

Kelly, Nora. *The Men of the Mounted.* Toronto: J.M. Dent & Sons, 1949.

Luciuk, Lubomir and Marco Carynnyk, eds. *Between Two Worlds: The Memoirs of Stanley Frolick.* Toronto: Multicultural History Society of Ontario, 1990.

Norton, Wayne and Naomi Miller, eds. *The Forgotten Side of the Border.* Kamloops: Plateau, 1998.

Ramsay, Bruce. *The Noble Cause: The Story of the United Mine Workers of America in Western Canada.* Calgary: District 18, United Mine Workers of America, 1990.

Rice, George S. *Accidents From Falls of Roof and Coal.* Washington: Miners' Circular 9, Bureau of Mines, U.S. Department of the Interior, 1912.

Seager, C. Allen. *F.H. Sherman of Fernie and the UMW of A 1902–1909: A Study in Western-Canadian Labour Radicalism.* Burnaby: Simon Fraser University, 1981.

Spatuk, Ann, ed. *Crowsnest and Its People.* Coleman: Crowsnest Pass Historical Society, 1979.

Steele, C. Frank. *Prairie Editor: The Life and Times of Buchanan of Lethbridge.* Toronto: Ryerson Press, 1961.

Sunahara, Ann Gomer. *The Politics of Racism: The Uprooting of Japanese Canadians During the Second World War.* Toronto: James Lorimer and Company, 1981.